POPPING WITH POWER

AUTHORS

Carol Bland

Betty Cordel

Helen Crossley

Susan Dixon

Sean Greene

Loretta Hill

Helen Rayfield

Anne Rudig

Ann Wiebe

Gina Wiens

Nancy Williams

Dave Youngs

EDITORS

Betty Cordel

Judith Hillen

Ann Wiebe

ILLUSTRATORS

Reneé Mason

Margo Pocock

Brenda Richmond

ORIGINAL PROJECT FACILITATOR

Jeri Starkweather

This book contains materials developed by the AIMS Education Foundation. **AIMS** (**A**ctivities **I**ntegrating **M**athematics and **S**cience) began in 1981 with a grant from the National Science Foundation. The non-profit AIMS Education Foundation publishes hands-on instructional materials (books and the monthly *AIMS* magazine) that integrate curricular disciplines such as mathematics, science, language arts, and social studies. The Foundation sponsors a national program of professional development through which educators may gain both an understanding of the AIMS philosophy and expertise in teaching by integrated, hands-on methods.

ISBN **1-881431-68-1**

Printed in the United States of America

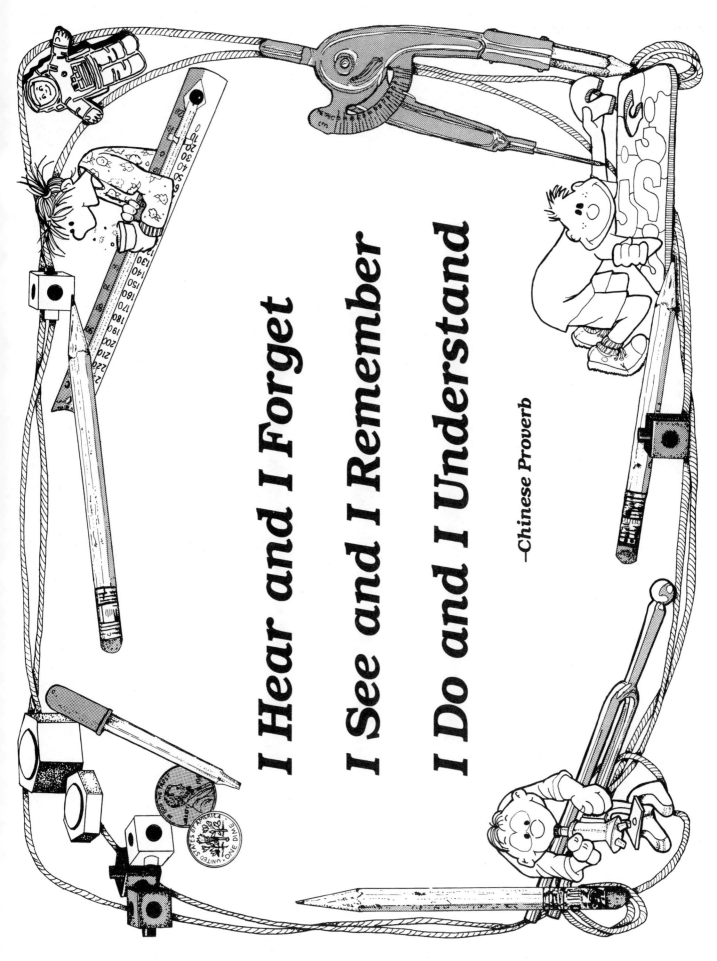

I Hear and I Forget

I See and I Remember

I Do and I Understand

—*Chinese Proverb*

Table of Contents

MATH	Estimation	Counting	Number sense and numeration	Measurement	Whole number operations	Decimals	Orders	Geometry and spatial sense	Averages	Graphs	Patterns	Problem solving	Relationships
A First-Class Job				X									
Fulcrums on the Move										X			X
Level the Lever	X			X	X						X	X	
Taking It Easy		X											
Working Together		X								X			
Metal Detectors													
Swinging Bears, Part I		X		X						X			
Swinging Bears, Part II		X		X						X			
Have a Ball	X			X			X						
On the Rebound	X			X						X	X		
From the Ground Up	X			X			X						
Ball on a Roll	X			X					X	X			
Polar Brrrs				X	X					X			
Cartons 'n Cotton				X	X								
Tints and Temps				X			X			X			
Why Be a Hot Head?				X	X					X			
Curly Cue													
Puff Mobiles	X			X						X			
Wind Rollers	X			X									
Blow Ye Winds	X			X				X					
Meter Readers	X		X		X					X			
Watts Going On			X		X		X						
Lighten Up	X	X			X	X				X			
Static Magic				X			X			X			
Slip, Siding Away				X									
Catapults				X	X				X				

PROCESSES

	Predicting	Observing	Classifying	Collecting/recording data	Comparing and contrasting	Identifying/controlling variables	Hypothesizing	Interpreting data	Inferring	Generalizing	Applying
A First-Class Job	X	X		X	X			X		X	X
Fulcrums on the Move	X	X		X	X			X		X	
Level the Lever	X	X		X	X			X		X	X
Taking It Easy		X	X	X				X			
Working Together		X		X				X			
Metal Detectors		X	X		X				X	X	
Swinging Bears, Part I		X		X	X		X			X	X
Swinging Bears, Part II		X		X	X		X			X	X
Have a Ball		X		X	X	X	X			X	
On the Rebound	X	X		X	X					X	
From the Ground Up	X	X		X	X	X		X			
Ball on a Roll		X		X	X	X				X	
Polar Brrrs		X		X							X
Cartons 'n Cotton		X		X				X			
Tints and Temps	X	X		X	X	X				X	X
Why Be a Hot Head?	X	X		X				X			X
Curly Cue		X			X			X			
Puff Mobiles		X		X	X	X	X				
Wind Rollers		X		X	X						X
Blow Ye Winds		X		X	X	X				X	
Meter Readers		X		X	X	X				X	X
Watts Going On	X	X		X	X						X
Lighten Up		X		X	X			X		X	X
Static Magic	X	X		X				X			
Slip, Siding Away	X	X		X	X					X	
Catapults		X		X							

A First-Class Job

Topic
First-class levers

Key Question
What happens when the position of the fulcrum on a first-class lever is changed?

Focus
Students will discover that less force is required to lift an object with a first-class lever when the fulcrum is placed closer to the resistance (load). They will also find that moving the fulcrum closer to the resistance means the load cannot be lifted as high.

Guiding Documents
NCTM Standards
- *Estimate, make, and use measurements to describe and compare phenomena*
- *Make inferences and convincing arguments that are based on data analysis*

Project 2061 Benchmarks
- *People can often learn about things around them by just observing those things carefully, but sometimes they can learn more by doing something to the things and noting what happens.*
- *People can use objects and ways of doing things to solve problems.*
- *Seeing how a model works after changes are made to it may suggest how the real thing would work if the same were done to it.*

Mathematics
Measuring
    linear

Science
Physical science
    simple machines
        first-class levers

Integrated Processes
Observing
Predicting
Comparing and contrasting
Collecting and recording data
Interpreting data
Generalizing
Applying

Materials
2" x 6" x 8' board
Concrete block
Box filled with 10 reams of paper
2 meter sticks

Background Information
Simple machines help us do work by trading force and distance. They do NOT lessen the work because you can't get something for nothing. With simple machines, there is a trade-off, force for distance, or distance for force.

The lever is one type of simple machine. All lever systems are made up of four components: the rod, which is called the *lever*, the pivot point, which is called the *fulcrum*; the point where the *effort force* is applied; and the point where the *resistance force*, or load, is located. (*Resistance* and *load* are used interchangeably. Use the term that suits your situation.) The *class* of the lever is determined by the center element in the arrangement of the effort, resistance, and fulcrum along the lever.

In a first-class lever, the fulcrum is located between the resistance and the effort. The closer the fulcrum is to the resistance, the less effort it takes to lift it. The trade-off occurs in that the distance the resistance can be lifted is decreased. You can't get something for nothing. In this activity, students should also notice that the effort always moves in the opposite direction of the load. If they push down on the lever, the load goes up.

Mechanical advantage is a comparison of the resistance force (r) to the effort force (e). The ratio (r/e) informs us about the number of times the machine multiplies the force used to do the job. The term and the quantification of mechanical advantage are not introduced in this lesson; however, it is hoped that students will begin to build a conceptual understanding of the mechanical advantage of a first-class lever.

Management
1. This activity is intended for whole-class exploration.
2. Allow enough time that all students can experience the difference in the amount of effort required to lift the paper in the box with the fulcrum placed at various distances from the resistance.
3. Use proper vocabulary throughout the lesson: resistance (load), fulcrum, effort.
4. To move the fulcrum, it is easiest to pull or push on the board, leaving the concrete block in one position.

Procedure

Part One

1. Allow several students to try to lift the box of paper. (Caution them not to strain their backs; the purpose is to discover that the box is heavy.)

2. Set up the first-class lever system as illustrated. Put the box of paper on one end of the lever. With the fulcrum near the effort, let students try to lift the box of paper using the lever.

3. Ask them what they think can be done to decrease the amount of effort that is required to lift the box. Try various solutions.

4. If students don't suggest moving the fulcrum, encourage them to design a systematic way in which to investigate that variable.

5. Allow time for all students to sense the differences in the amount of effort required to lift the box with the fulcrum at various positions.

6. Ask them to make a generalization.

7. Discuss other observations such as: the resistance moved the opposite direction as our effort; when we moved the fulcrum closer to the resistance, it was easier to lift the box; we couldn't lift the box very high when the fulcrum was close to it.

8. Discuss the meaning of the following: Simple machines make a trade-off between force and distance.

9. Find other heavy objects in the classroom to lift using the lever.

10. Have students illustrate a lever that requires a great deal of force to lift the resistance. Ask them to use their own words to tell why it requires a lot of force.

Part Two

1. Inform students that they are going to do the same procedures again, but this time they will measure how high they can lift the load and how far they have to push down the effort. Ask them to discuss as a class how to organize the procedure and to make a record on their activity pages of their results.

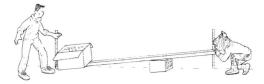

2. Follow the class plan to collect the data.

3. Direct small student groups to get together to discuss a generalization of their interpretation of the data they just collected.

Part Three

1. Inform the students that instead of them applying force to the lever to lift the paper, they will be using reams of paper to lift reams of paper.

2. Open the box of paper and take out five reams to represent the resistance. Ask a student to place them on one end of the lever. Tell the students that the other five reams will be used for the effort.

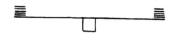

3. Ask them to predict where the fulcrum should be placed so that the five-ream effort can lift the five-ream resistance. Have them explain their reasoning. Let them try their suggestions.

4. Ask what would need to be done if only four reams of effort were used. Again ask for an explanation. Try their suggestions.

5. Follow the same procedure for three, two, and one ream of effort.

Discussion

1. Why is this lever called a first-class lever? [the fulcrum is located between the effort and the resistance.]

2. Does the fulcrum have to be located halfway between the effort and the resistance in order to be a first-class lever? Explain.

3. What is the advantage of using a lever? [It allows us to use less force to lift an object.]

4. What is the cost of using less effort? [Distance, you can't lift the resistance as high.]

5. What are some real-world examples of a first-class lever?

6. When would you use a first-class lever?

Extensions

1. Do other AIMS activities such as *Fulcrums on the Move*, *Level the Lever*, and *Beams Over Board*.

2. Use the playground seesaw to investigate first-class levers.

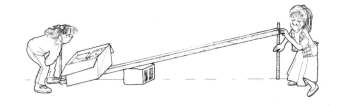

A First-Class Job

Part One

Illustrate a lever that requires a lot of force to lift the box of paper. Label the effort, fulcrum, and resistance.

Explain why this lever requires a lot of force to lift the paper.

How could you make this job easier?

A First-Class Job

Part Two

Make a record of the distance the paper moves and the distance the effort moves.

What conclusion(s) can you make from your data?

Part Three

Illustrate the levers that lift:

- 5 reams with 5 reams of effort

- 5 reams with 3 reams of effort

- 5 reams with 1 ream of effort

Fulcrums on the Move

Topic
First-class levers

Key Question
If you want to balance a first-class lever, what must you do when you decrease the amount of effort force?

Focus
Students will discover that in a first-class lever less effort force is required as the fulcrum is moved closer to the resistance force.

Guiding Documents
NCTM Standards
  • *Construct, read, and interpret displays of data*

Project 2061 Benchmarks
  • *People can often learn about things around them by just observing those things carefully, but sometimes they can learn more by doing something to the things and noting what happens.*
  • *People can use objects and ways of doing things to solve problems.*
  • *Seeing how a model works after changes are made to it may suggest how the real thing would work if the same were done to it.*
  • *Tables and graphs can show how values of one quantity are related to values of another.*
  • *Graphical display of numbers may make it possible to spot patterns that are not otherwise obvious, such as comparative size and trends.*

Math
Graphing
Relationships

Science
Physical science
  simple machines
    first-class levers

Integrated Processes
Observing
Predicting
Comparing and contrasting
Collecting and recording data
Interpreting data
Generalizing

Materials
Per group:
  48 Hex-a-Link (interlocking) cubes

pencil
masking tape
2 different colored crayons or markers

Background Information
  Simple machines help us do work by trading force and distance. They do NOT lessen the work because you can't get something for nothing. With simple machines, there is a trade-off, force for distance, or distance for force.
  The lever is one type of simple machine. All lever systems are made up of four components: the rod, which is called the *lever*, the pivot point, which is called the *fulcrum*; the point where the *effort force* is applied; and the point where the *resistance force*, or load, is located. (*Resistance* and *load* are used interchangeably. Use the term that suits your situation.) The *class* of the lever is determined by the center element in the arrangement of the effort, resistance (load), and fulcrum along the lever.
  In a first-class lever, the fulcrum is located between the resistance and the effort. The closer the fulcrum is to the resistance, the less effort it takes to lift it. The graphical representation in this activity helps students in making the generalization that when they want to use less effort force to lift an object with a first-class lever, the fulcrum needs to be moved closer to the resistance.

Management
  1. This activity is intended for student groups of three or four.
  2. A bag of 500 Hex-a-Link cubes will be sufficient for a class of 35 students. Each group should have 48 cubes. It may be easier for students if 12 Hex-a-Link cubes of one color are used for the resistance force and 12 cubes of another color are used for the effort force.
  3. Use proper vocabulary throughout the lesson: resistance (load), fulcrum, effort.
  4. It is assumed students will have done *A First-Class Job*.
  5. Copy one set of *Lever Labels* for each group to help clarify the positions on the lever.
  6. If the cubes of the lever don't hang down properly, suggest to the students that they may need to be twisted slightly. Give them time to tinker with the setup.

Procedure

1. Allow time for students to assemble the lever system. Have them position the labels (*Resistance, Fulcrum,* and *Effort*) as indicated in the illustration.

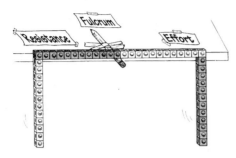

2. Review the generalizations made in the activity *A First-Class Job.*

3. Inform the students that the resistance force will remain the same (12 Hex-a-Link cubes) for this entire activity.

4. Ask students where they think the fulcrum will be located to balance a resistance of 12 Hex-a-Link cubes with an effort force of 12 Hex-a-Link cubes.

5. Direct the students to **lightly** draw a line on their activity sheet to indicate their prediction for the placement of the fulcrum.

6. Allow time for them to balance the lever. Have them draw a **darker** line to indicate the actual location of the fulcrum to balance the lever. Urge them to color the resistance arm one color and the effort arm the other color.

7. Guide them to remove two Hex-a-Link cubes from the effort force. Have them again draw a light line to indicate their prediction of where the fulcrum will be located to balance the lever.

8. Have them test their predictions, find and record the actual results, and color in the arms of the lever using the color scheme in the previous trial.

9. Urge the students to continue the procedure, reducing the effort force by two Hex-a-Link cubes each time.

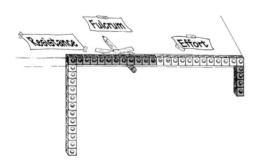

10. Ask them to write about the relationship between the number of cubes used for the effort force and the length of the effort arm.

Discussion

1. What did you learn in this activity?
2. How is this activity like the activity *A First-Class Job*?
3. What do you notice about the number of cubes on the resistance side of your graph? [The number of cubes stays the same.]
4. What do you notice about the number of cubes on the effort side of your graph? [The number of cubes gets smaller.]
5. What do you notice about the lengths of the bars on the effort side of your graph as the number of cubes gets smaller? [The lengths of the bars get longer.] What does this tell you?
6. When using a first-class lever, what do we have to do to use less effort to lift something? Explain how your data record supports this.

6

Fulcrums on the Move

  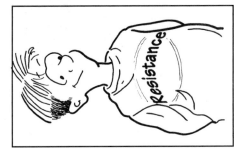

7

Fulcrums on the Move

Each square represents one cube. Draw a line to indicate the position of the fulcrum that balances the system. Color the resistance arm one color and the effort arm another color.

Resistance   Effort

12 [grid] 12
Describe the fulcrum's position:

12 [grid] 10
Describe the fulcrum's position:

12 [grid] 8
Describe the fulcrum's position:

12 [grid] 6
Describe the fulcrum's position:

12 [grid] 4
Describe the fulcrum's position:

12 [grid] 2
Describe the fulcrum's position:

On the back side of this paper, write what you know about the amount of effort needed to balance the lever and the position of the fulcrum.

Topic
First-class lever

Key Question
How can you balance a first-class lever?

Focus
Students will discover the mathematical pattern for balancing a first-class lever.

Guiding Documents
NCTM Standards
- *Verify and interpret results with respect to the original problem*
- *Make and use measurements in problems and everyday situations*
- *Represent and describe mathematical relationships*

Project 2061 Benchmarks
- *Measuring instruments can be used to gather accurate information for making scientific comparisons of objects and events and for designing and constructing things that will work properly.*

Math
Patterns
Measuring
Estimating
Computation
Problem solving

Science
Physical science
   simple machines
     first-class lever

Integrated Processes
Observing
Predicting
Comparing and contrasting
Collecting and recording data
Interpreting data
Generalizing
Applying

Materials
Per group:
   1 50-cm wooden strip (see *Management*)
   2 medium binder clips
   pencil
   tape
   distance measurer, included
   clay, optional (see *Management*)
   Unifix Cubes or Hex-a-link Cubes

Background Information
The scientific formula for determining work is force times distance (W = f x d). Simple machines help us do work by trading force for distance. They do NOT lessen the work; there is a trade-off, force for distance.

The lever is one type of simple machine. All lever systems are made up of four components: the rod, which is called the *lever*; the pivot point, which is called the *fulcrum*; the point where the *effort force* is applied; and the point where the *resistance force*, or load, is located. The *class* of the lever is determined by the center element in the arrangement of the effort, resistance, and fulcrum along the lever. In a first-class lever, the fulcrum is located between the resistance and the effort. A familiar example is a seesaw.

In order to balance a first-class lever, the forces (resistance and effort) on either side of the fulcrum must be evenly distributed. This means that the force times the distance on one side of the fulcrum must equal the force times the distance on the other side. (Non-customary measuring units will be used in this activity, Unifix Cubes and corresponding distance units.)

Resistance Side
3 Unifix Cubes
Distance from fulcrum
  = 10 distance units

Effort Side
3 Unifix Cubes
Distance from fulcrum
  = 10 distance units

$$3 \times 10 = 3 \times 10$$

Resistance Side
6 Unifix Cubes
Distance from fulcrum
  = 12 distance units

Effort Side
6 Unifix Cubes
Distance from fulcrum =
  10 distance units and
  1 Unifix Cube at 12
  distance units

$$6 \times 12 = (6 \times 10) + (1 \times 12)$$

In both examples, the force times the distance on the resistance side of the fulcrum equals the force times the distance on the effort side of the fulcrum. The lever system is therefore balanced!

Management

1. Options for the wooden strips:
   - Purchase lattice strips or laths from lumber supplies.
   - Spacer strips are often free for the asking from lumber supplies. These are the wooden strips placed between larger boards to allow for air circulation. They are quite rough and will need to be cut and sanded. To engage students in the preparation of materials, an area can be set up a couple of days ahead of time where students saw the spacer strips to the appropriate length and then sand their surfaces.
   - Rip a 2" x 4" board into pieces approximately 2" wide x 1/4" deep x 22 1/4" long. Students in high school Industrial Technology classes may be willing to do this for you.
   - Wooden meter sticks can be cut in half.
2. Copy the included distance measurer onto colored paper. Laminate them if possible to help preserve them.
3. Students should not spend a great deal of time with the initial balancing of their levers. Clay or paper clips can be used to balance them, but they must remain in place for the entire activity.
4. Assemble a lever system beforehand so that students can use the model for their constructions.
   a. Cut out the distance measurer. Tape it together so that the dark side and light side meet and measure out from the center.

`12 11 10 9 8 7 6 5 4 3 2 1 0 1 2 3 4 5 6 7 8 9 10 11 12`

   b. Tape the distance measurer on top of the wooden strip.
   c. Attach two binder clips to the center of the lever over the zero on the distance measurer.
   d. Use tape to secure a pencil to the table or desk. Make sure the eraser end of the pencil extends over the edge.
   e. Hang the lever on the pencil end using the loops of the binder clips.

If necessary, balance the lever with pieces of clay or paper clips.
5. Have students work in groups of four.
6. For the sake of consistency, have students place the Unifix Cubes so they are centered over the measur-

ing marks. The Unifix Cube trains should be placed perpendicular to the distance measurer; vertically placed Unifix Cube trains will easily topple when the lever is not balanced.
7. If appropriate for your students, use the terms *effort force*, *effort arm*, *resistance force*, and *resistance arm*. (*Resistance* and *load* are interchangeable terms. Use the term that fits your situation.)

Procedure

1. Ask the *Key Question*.
2. Allow time for students to assemble their lever systems. Have the students make certain the levers are level (balanced).
3. Distribute Unifix Cubes. Ask students to place one Unifix Cube on the 10 distance mark on the dark side of the measuring tape. Discuss what happens to the lever.
4. Ask students how they could "Level the Lever." Give them some free exploration time to work at leveling the lever with various groupings of Unifix Cubes.
5. Invite students to share what they have discovered.
6. Distribute the first two activity pages. Urge students to apply what they have learned in their free exploration time to the tasks on the page. Encourage them to guess and test. If their predictions don't match the actual results, have them continue to work at the tasks until they begin to see the relationship of force times distance.
7. Once students understand the relationship, have them make up several situations to challenge other groups using the activity sheet *Designing Your Own Work*. For example: On one side of the lever, 3 Unifix Cubes at 10 distance units and 2 Unifix Cubes at 12 distance units. Establish the rule that the other groups cannot duplicate the scenario on the opposite side of the lever; they must use a different combination of force and distance units. (The other groups may choose to place 4 Unifix Cubes at 10 cm or 2 Unifix Cubes at 7 cm.) The students will cut out and glue the *Cube Patterns* on top of the lever's illustrated distance measures. The group that is solving the problems will also cut out and glue the *Cube Patterns* on the opposite arm of the illustrated lever.
8. Once the groups have solved each other's scenarios, ask them to share results with the entire class. Urge students to use their multiplication skills to determine if the various solutions will work, then try them out to verify.
9. Have students generalize the pattern for balancing the lever. [force times distance on one side must equal force times distance on the other side]

Discussion

1. How did you determine whether the lever was balanced?
2. When did you notice there was a pattern to the numbers that you could use to balance the lever?
3. Did anyone try other patterns that you later found wouldn't apply to all situations? If so, what were they? Why didn't they work?
4. How does what you learned in this activity apply to balancing a seesaw?
5. What was your favorite solution to another group's problem? Why did you like it?

Extensions

1. Use one object of known weight and balance an object of an unknown weight on the lever. Challenge students to determine the unknown weight.
2. Do *Beams Over Board* using LEGO® elements from the AIMS publication *Brick Layers*.
3. Apply the lesson of the first-class lever to mobiles. Do *Hanging in the Balance* (*AIMS*, Volume 9, Number 10).

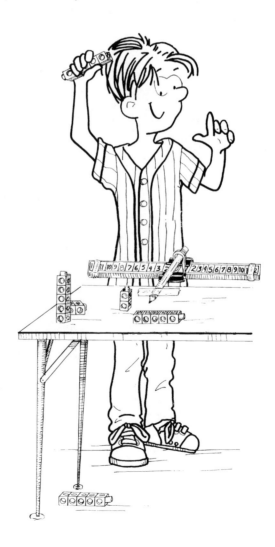

Level the Lever

Distance Measurer

Build the lever. Place the number of cubes at the distance shown in the table. Balance the lever by placing cubes on the other side of the lever. Record the distance.

Tell two more ways to balance one cube at distance 12.

Number of Cubes	Distance Units	Number of Cubes	Distance Units
1	10	1	
1	10	2	
1	6	1	
1	6	2	
1	6	3	
1	12	1	
1	12	2	
1	12	3	
1	12	4	

How can you balance two cubes at distance 10?

Write two number sentences to show that you understand the pattern that balances the lever.

Level the Lever

Now solve these problems.

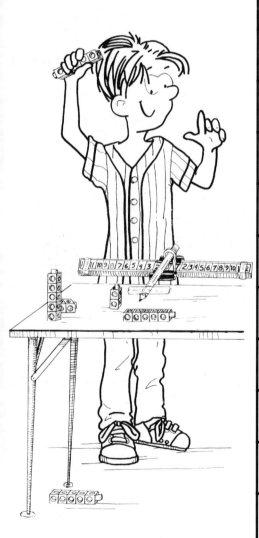

Number of Cubes	Distance Units	Number of Cubes	Distance Units
2	8		4
3	8		6
3	8		3
4	9	3	
6	4		2
4	6	2	
4	5		10
1 2	6 8	2	
2 2	5 7	3	
	3	2 2	5 7
4		2 2	3 11
		1 2	5 10

Level the Lever

- Cut out some cubes from the pattern below.

- Arrange and glue them on the dark side of each of the four levers.

- Trade papers with another group.

- Cut out and glue cube patterns to the light side of the levers to balance them.

- Write the number sentences that balance the levers.

Designing Your Own Work

Design your own lever balancing problems for another group to solve.

| 12 | 11 | 10 | 9 | 8 | 7 | 6 | 5 | 4 | 3 | 2 | 1 | 1 | 2 | 3 | 4 | 5 | 6 | 7 | 8 | 9 | 10 | 11 | 12 |

| 12 | 11 | 10 | 9 | 8 | 7 | 6 | 5 | 4 | 3 | 2 | 1 | 1 | 2 | 3 | 4 | 5 | 6 | 7 | 8 | 9 | 10 | 11 | 12 |

| 12 | 11 | 10 | 9 | 8 | 7 | 6 | 5 | 4 | 3 | 2 | 1 | 1 | 2 | 3 | 4 | 5 | 6 | 7 | 8 | 9 | 10 | 11 | 12 |

| 12 | 11 | 10 | 9 | 8 | 7 | 6 | 5 | 4 | 3 | 2 | 1 | 1 | 2 | 3 | 4 | 5 | 6 | 7 | 8 | 9 | 10 | 11 | 12 |

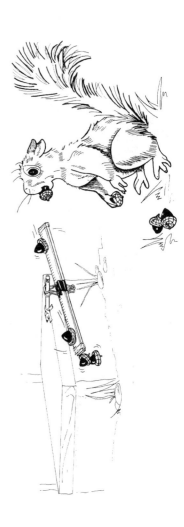

Cube Patterns

Take It Easy

I. Topic Area
Simple Machines

II. Introductory Statement
Students will learn how the six simple machines work.

III. Math Skills
a. Counting

Science Processes
a. Observing and classifying
b. Gathering and recording data
c. Interpreting data

IV. Materials
1 class set of the 6 simple machines.
"Suggestions"

inclined plane	**pulley**
slide	drapes
various ramps	flagpoles

wedge	**wheel and axle**
knives	doorknobs
forks	toy cars
sewing needles	fishing pole

lever	**gear wheels**
hammer	hand drill
bottle opener	hand mixer
	bicycle

1 class set of task cards
"Examples"

inclined plane
1. Roll the toy car down the ramp.

wedge (sewing needle)
1. Stick the needle into the cloth.
2. Pull the needle out.

lever (hammer)
1. Hammer the nail into the block of wood.
2. Use the hammer to pull the nail out.

pulley (drapes)
1. Use the cords to close the drapes.
2. Now open the drapes.

wheel and axle (doorknob)
1. Open the door using the knob.
2. Close the door with the knob.

gear wheels (hand drill)
1. Turn the handle of the drill so the point goes into the block of wood.
2. Pull the drill out of the wood.

materials to perform each task
chart paper

V. Key Question
Why do machines make your life easier?

VI. Background Information
There are six simple machines.
1. Inclined plane: A slanted surface, which is sometimes called a ramp.
2. Wedge: Two inclined planes joined together to form a sharp edge.
3. Lever: A bar resting on a turning point (fulcrum).
4. Pulley: A wheel with a rope moving around it.
5. Wheel and axle: A wheel that turns on a rod.
6. Gear wheel: A wheel with teeth.

VII. Management
1. A class period of 50-60 minutes.
2. Class divided into six working groups.
3. Set up a station for each machine.
4. Groups will rotate through the stations.

VIII. Procedure
1. Gather the six simple machines.
2. Prepare a task card for each machine.
3. Set up the six stations.
4. Prepare a chart size copy of the student worksheet.

IX. What the Students Will Do
Do the following at each station:
1. Record the name of the object.
2. Perform the job on the task card.
3. Record the job each object does.

Total Class
4. Record data from each group on class chart.

X. Discussion
1. Discuss how and why each simple machine works. (During discussion students will fill in information on student worksheet.)
2. How many simple machines do you see around the classroom (desk, playground, etc.)?
3. What are some of the machines you found?
4. Which kind of simple machine did you find the most of?
5. Which kind of simple machine was hardest to find?
6. How did we discover that simple machines make tasks in life easier?

XI. Extensions
1. Discover what parts of your body are simple machines.
2. Show "Donald Duck in Math Magic Land" (film-part IV)
3. Check your home for simple machines.

XII. Curriculum Coordinates
1. Write a poem about three simple machines.
2. Write a paragraph comparing and contrasting two simple machines.
3. Draw a picture of some simple machines in your own bedroom.
4. Do some physical exercise using the various portions of your body that are simple machines.

Take It Easy

NAME _____

1. GO TO EACH STATION.

2. OBSERVE THE MACHINE - READ THE TASK CARD - DO THE JOB.

3. RECORD THE OBJECT AND JOB AT EACH STATION ON THIS WORKSHEET.

4. CLASS DISCUSSION WILL FOLLOW.

STATION	OBJECT	JOB	HOW IT WORKS	NAME OF MACHINE
1				
2				
3				
4				
5				
6				

NAME _____

Take It Easy

MACHINE	QUANTITY	EXAMPLES	
INCLINED PLANE		A._____ B._____ C._____	D._____ E._____ F._____
WEDGE		A._____ B._____ C._____	D._____ E._____ F._____
LEVER		A._____ B._____ C._____	D._____ E._____ F._____
PULLEY		A._____ B._____ C._____	D._____ E._____ F._____
WHEEL AND AXLE		A._____ B._____ C._____	D._____ E._____ F._____
GEAR WHEELS		A._____ B._____ C._____	D._____ E._____ F._____

I. **Topic Area**

Compound Machines

II. **Introductory Statement**

Students will learn about compound machines.

III. **Math Skills**

a. Counting
b. Graphing

Science Processes

a. Gathering and recording data
b. Interpreting data
c. Observing and classifying

IV. **Materials**

chart paper
samples of compound machines
"Suggestions"
  bicycle
  hand drill
  scissors

V. **Key Question**

How many compound machines did you see on your way to school today?

VI. **Background Information**

1. The knowledge of simple machines would be prerequisite learning.
2. A compound machine is made up of two or more simple machines to do a job.

   Examples:
   Bicycle—wheel and axle, inclined plane, gear wheels
   Hand drill—wheel and axle, wedge shapes
   Scissors—wedges; two levers joined at a fulcrum

VII. **Management**

1. A class period of 50-60 minutes with an optional 30 minute independent follow-up.
2. Divide the students into small groups with a leader.
3. The leader should be responsible and will have to study the information ahead of time.
4. Groups will visit different areas at school looking for compound machines (playground, office, kitchen, classroom).

VIII. **Procedure**

1. Prepare charts.
2. Arrange tour of school kitchen and office.
3. Get some examples of compound machines.

IX. **What the Students Will Do**

1. Review the six simple machines and how they work.
2. Look at examples of compound machines.
3. Tour the office, classroom, kitchen, and playground. Look for examples of compound machines.
4. Record your findings on the student worksheet.
5. Transfer your findings to a bar graph.

X. **Discussion**

1. What compound machines did you find?
2. What simple machines were in each of the compound machines you found?
3. In which area of the school were the most compound machines found?
4. Why do you think this is so?
5. Can you think of ten different compound machines you'd find in your home?

XI. **Extensions**

1. Take a more complicated machine (radio, engine) and try to identify the simple machines involved.
2. Actually build your compound machine invention for a class demonstration.
3. As a class construct a compound machine to help the students do some simple task more efficiently.
4. Try to create your own compound machine to do an inventive task. Examples: shoelace tier, face-washing machine, bubble gum blower, garbage taker outer, exercising machine for fish.
5. Label your own invention with the simple machines that make it up.

XII. **Curriculum Coordinates**

1. Write a story about your compound machine invention.
2. While at P.E. try to identify the machines you are playing on.
3. Construct a toothpick model of your compound machine invention.

Working Together

COMPOUND MACHINES I FOUND:

CLASSROOM	OFFICE	PLAYGROUND	KITCHEN

NAME _____

Working Together

	CLASSROOM	OFFICE	PLAYGROUND	KITCHEN
14				
13				
12				
11				
10				
9				
8				
7				
6				
5				
4				
3				
2				
1				
0				

Metal Detector

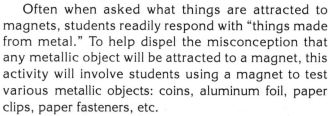

Topic
Magnetic attraction

Key Question
What do the things that are attracted to our magnet have in common?

Focus
Students will find that some metals, not all, are attracted to a magnet.

Guiding Document
Project 2061 Benchmark
- *Without touching them, a magnet pulls on all things made of iron and either pushes or pulls on other magnets.*

Science
Physical science
    magnetism

Integrated Processes
Observing
Classifying
Comparing and contrasting
Generalizing
Inferring

Materials
Large paper grocery bags
Sand
Assorted metallic objects (see *Management 1*)
Ring magnets or cow magnets (see Management 6)
Plastic wrap
Pencils or dowels

Background Information
Metals such as iron, cobalt, or nickel (also rare earths and alloys such as steel) are magnetic materials. This means that they can be attracted by magnets, and in some cases even made into magnets. In the atomic structure of these metals, clusters of atoms are aligned magnetically in small areas called magnetic domains. In most magnetic materials, these domains are arranged randomly and they produce no overall magnetic field. However, in the presence of a strong external magnetic field, many of these domains become aligned allowing a magnet to attract an object made of these metals. If this alignment is permanent, the object becomes a magnet.

Often when asked what things are attracted to magnets, students readily respond with "things made from metal." To help dispel the misconception that any metallic object will be attracted to a magnet, this activity will involve students using a magnet to test various metallic objects: coins, aluminum foil, paper clips, paper fasteners, etc.

All science begins with observations, whether qualitative or quantitative. Observations naturally lead to the making of inferences and predictions.

An inference is an attempt to explain an observation of an event or phenomena. In this activity, students will attempt to explain which metals are attracted to magnets and which are not. Inferences may or may not be scientifically accurate. It is important to probe for the reasoning that underlies the inferences. For instance, students may infer that tin objects are attracted to a magnet because they have observed that the magnet sticks to a "tin can." In reality, the magnet is attracted to the steel that is in the can. In order to alter this incorrect inference, students would need to experience a magnet's reaction to tin.

For further information about magnets and magnetism, refer to *Science Information* in the AIMS publication *Mostly Magnets.*

Management
1. Prior to the activity ask students to bring in small metallic objects. Also have on hand things like paper clips, paper fasteners, aluminum foil, coins, pen caps with metal clasps, etc. If possible, include some magnetic tape from an old audio cassette.
2. Have students work in groups of four.
3. Each group will need a large grocery bag with the top folded down or cut off so that they have a container that is about eight centimeters deep.
4. Add sand to the bag to a depth of about two centimeters. This is their *Bag of Beach* which they will comb for magnetic tresure.
5. The magnets used in this activity are wrapped in plastic wrap for ease of cleaning the iron particles from the sand that are attracted to them.
6. Ring magnets or cow magnets will both work for this activity. Use the stronger cow magnets if they are available. (Both ring magnets and cow magnets can be ordered from AIMS.)

22

Procedure

1. Ask the students to list the metallic objects their group members brought in. Include, if necessary, items from those you have assembled. Have them

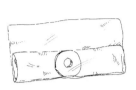 

predict which items in their collection will be attracted to a magnet.

2. Discuss the similarities of the items students have predicted will be attracted to the magnet.
3. Direct them to bury the objects in the *Bag of Beach*.
4. Inform them that they will be making a "metal detector" out of a magnet to test their predictions.
5. Have students wrap the magnet in a piece of plastic wrap and tie the ends of the plastic wrap around the end of a pencil or dowel.
6. Direct them to hold the metal detector near the surface of the sand to see which objects are attracted to the magnet. (For weaker magnets it may be necessary to either use less sand or to run the metal detector through the sand.)
7. Ask students to classify their objects into two groups, attracted by the magnet and not attracted by the magnet.
8. Have them compare their results to their predictions.
9. Discuss the similarities of the objects that were attracted to the magnets.
10. Encourage students to form a generalization about objects that are attracted to magnets. [Not all metallic objects are attracted to magnets; Only some metals are attracted to magnets.]
11. Have students bury their objects in the sand once again and trade bags with another group. Encourage them to use the metal detector once again. Ask if they were surprised about their results this time.

Discussion

1. When you made your first predictions, what kind of objects did you think would be attracted to the magnet?
2. Were you surprised or puzzled by your results?
3. How did you have to change what you thought about magnetic objects?
4. Would our metal detector be good for finding hidden treasure? Explain.
5. Why shouldn't our device be called a metal detector? [It doesn't detect all metals; it only attracts objects with magnetic properties.]
6. What other objects would you like to try?
7. Why did we wrap the magnet in plastic wrap? [to make the iron particles from the sand easier to remove from the magnet]
8. To find out what metals are attracted to a magnet, what other tests could we do?

Extension

See *Mostly Magnets* for more activities dealing with magnets and magnetism.

Metal Detector

1. Wrap the magnet 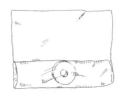 in plastic wrap.

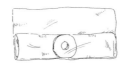

2. Tie the ends of the plastic wrap around the end of your pencil.

List the objects hidden in your "bag of beach."

Check (✓) the ones you predict will be attracted by the magnet.

Now comb the beach for the treasure.

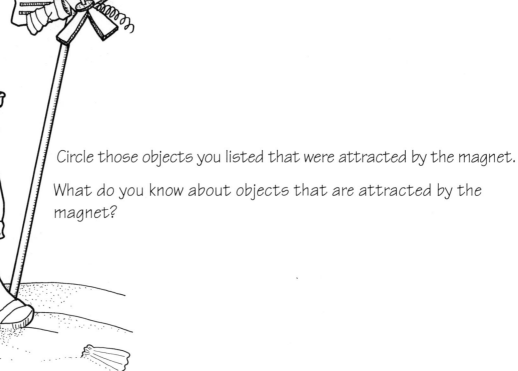

Circle those objects you listed that were attracted by the magnet.

What do you know about objects that are attracted by the magnet?

Part One

Topic
Pendulums

Key Question
How many cycles will your pendulum make in 30 seconds?

Focus
Students will discover the relationship between pendulum length and frequency by using pendulums to make a real graph.

Guiding Documents
NCTM Standards
- *Understand the attributes of length, capacity, weight, mass, area, volume, time, temperature, and angle*
- *Link conceptual and procedural knowledge*
- *Construct, read, and interpret displays of data*

Project 2061 Benchmarks
- *The earth's gravity pulls any object toward it without touching it.*
- *Mathematics is the study of many kinds of patterns, including numbers and shapes and operations on them. Sometimes patterns are studied because they help to explain how the world works or how to solve practical problems, sometimes because they are interesting in themselves.*

Math
Counting
Measurement
Graphs

Science
Physical science
    pendulums
    gravity

Integrated Processes
Observing
Collecting and recording data
Comparing
Hypothesizing
Generalizing
Applying

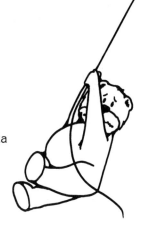

Materials
For the class:
    number line from 1-50
    clock with second hand
    pushpins

For each group:
    2 Friendly Bears or Teddy Bear Counters
    2 pieces of string (see *Management 2*)
    masking tape

Background Information
    Pendulums were first studied in depth by Galileo in the sixteenth century. He discovered the regular motion of pendulums while watching a lamp swaying in a cathedral in Pisa. Using his pulse to time the swing of the lamp, he found that although the arc through which the lamp swung steadily got smaller, its *period* (the time it took to make one complete out and back cycle) remained constant. Fascinated by this experience, Galileo began to study pendulums. By careful observation and experimentation, he made several important discoveries. For one thing, he found that the length of the pendulum determined its *frequency* (cycles per minute): the longer the pendulum, the greater its period, and consequently, the lower its frequency. Secondly, he found that in order to predict the frequency of a pendulum, its length must be measured from its pivot point to its center of mass. Thirdly, he discovered that varying the mass of a pendulum while keeping its length the same did not change its frequency. These discoveries were very important, for they led to the first accurate clock (built shortly after his death) and helped Newton formulate his laws of motion.

Management
1. This activity is designed to be done without a student page. Instead, students help create a real graph using their pendulums.
2. Strings for the pendulums need to be prepared ahead of time. Cut strings to random lengths between 10 and 70 cm. Make sure there is a good distribution of short, medium, and long strings. Each group needs two strings.
3. If Friendly Bears or Teddy Bear Counters are not available, pennies, or other small, uniform objects can be substituted as pendulum bobs.
4. Students should work in groups of three or four.
5. The ideal place for the pendulum graph is a bulletin board with a number line from 1 to 50 across the

top. If this is not available, write the numbers 1 to 50 across the top of a chalkboard. For pendulums longer than 10 cm, the number of cycles in 30 seconds falls within this range.

Procedure
1. Give each group two strings, two bears, and some tape.
2. Show students how to make a pendulum by taping the bear to one end of the string and tying a knot (not a loop) in the other end.
3. Once the pendulums are made, have each group pick one pendulum and tell them they will count how many cycles (one complete out and back swing is a cycle) it makes in 30 seconds. Explain that the pendulum bobs should be started from about a 45-degree angle and that it is important that the top of the pendulum (the knot) be held as steady as possible.
4. After students have finished counting the number of cycles, use pushpins (poked through the knot) to help them hang the pendulums under the appropriate numbers on the number line (see illustration).
5. Tell students to look at the graph and use it to predict how many cycles their second pendulums will make in 30 seconds.
6. After groups have made their predictions, they should count the cycles to see how close their predictions were and then hang the second pendulums on the graph.
7. Discuss the pendulum activity.

Discussion
1. What generalizations about pendulums can you make?
2. What patterns do you see in the graph?
3. How did the graph help you in making your prediction for the second pendulum?
4. How long would you need to make a pendulum to give you ten cycles in 30 seconds? How could you find out?
5. Why do you think pendulums are used in grandfather clocks and cuckoo clocks?
6. The weight at the bottom of a grandfather clock's pendulum can be moved up and down. If the clock is running slow, which way should the weight be moved? Why?
7. What other pendulum questions would you like to explore?

Extensions
1. Show the relationship between the pendulum graph and the acceleration due to gravity by gently throwing a ball horizontally in front of the graph, starting at the level of the shortest pendulum. If the ball is tossed at the right speed, its downward curve will match the curve made by the pendulums.
2. Have students devise experiments to answer some of their pendulum questions.
3. Go outside and observe the pendulum motion of the playground swing.

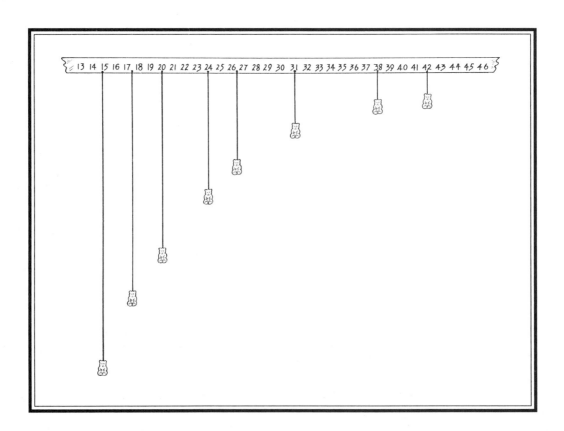

Part Two

Topic
Pendulums

Focus
Students will discover the relationship between pendulum length and frequency by experimenting with pendulums of various lengths and constructing a line graph.

Guiding Documents
NCTM Standards
- *Understand the attributes of length, capacity, weight, mass, area, volume, time, temperature, and angle*
- *Link conceptual and procedural knowledge*
- *Construct, read, and interpret displays of data*

Project 2061 Benchmarks
- *The earth's gravity pulls any object toward it without touching it.*
- *Mathematics is the study of many kinds of patterns, including numbers and shapes and operations on them. Sometimes patterns are studied because they help to explain how the world works or how to solve practical problems, sometimes because they are interesting in themselves.*

Math
Counting
Measurement
Graphs

Science
Physical science
   pendulums
   gravity

Integrated Processes
Observing
Collecting and recording data
Comparing
Hypothesizing
Generalizing
Applying

Materials
For the class:
  clock or watch with second hand

For each group:
  1 Friendly Bear or Teddy Bear Counter
  one 65-100 cm piece of string
  masking tape
  meter stick or meter tape

Key Question
How does the length of a pendulum affect its frequency?

Background Information
The frequency of a pendulum is a function of its length: the shorter the length, the higher the frequency. See *Background Information* in *Swinging Bears, Part I.*

Management
1. *Swinging Bears, Part I* should be done before this activity.
2. Students need to know how to construct a line graph in order to do this activity.
3. Students should know the terms *cycle* (one complete out and back motion), *period* (the time for one cycle), and *frequency* (the number of cycles per minute).
4. Cooperative teams of three or four can collect data, but each student should have a copy of the activity sheet and graph page to record the team's results. Students can take turns with the various tasks: measuring the pendulum lengths, holding the pendulum, counting the cycles, and timing. If the string is held against the bottom edge of a table or desk, more accurate results are possible.
5. One string will be used for the various pendulum lengths. The length of each pendulum is measured from the pivot point to its center of mass (approximately the middle of the pendulum bobs). A meter stick should be used to carefully measure this distance for each different length. The loose end of the string should be placed on top of the table where it will be out of the way.

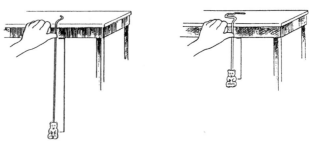

Changing Pendulum Lengths

© 1996 AIMS Education Foundation

6. The activity sheet and graph page are open-ended, allowing the teacher to choose the number and range of pendulum lengths to be tested.

7. Before doing the activity, decide on the pendulum lengths that will be used. A suggested range of lengths is 10 to 60 cm, with 10 cm increments. Longer pendulums can be used, but they will be too long to suspend from a table or desk. Pendulums shorter than 10 cm have high frequencies and are difficult to count.

8. Since the shorter pendulums have higher frequencies and are more difficult to count, it might be helpful to start the activity with the longer pendulums and finish with the shorter ones. This gives students practice in counting the cycles of the lower frequency pendulums first. Another way to make the counting easier is to time the pendulums for 30 seconds and double this number.

9. In making the graph, students should label the x-axis (horizontal) with the pendulum length (independent variable) and the y-axis (vertical) with the frequency (dependent variable). Scales for each axis should be chosen so that the maximum amount of graphing area is used.

Procedure

1. Discuss the *Key Question*.
2. Divide students into groups.
3. Give each group a string, tape, bear, and meter stick, as well as activity sheets for each student.
4. Each group should tape their bear (or other small object) securely to one end of the string.
5. Have students record the pendulum lengths that will be tested in the appropriate boxes on the activity sheet.
6. Explain the process for measuring the various pendulum lengths and finding their frequencies.
7. Students should do the activity and record their results on the activity sheet.
8. After discussing the results, have students write their observations and conclusions on the activity sheet.
9. Students should use the data collected to construct a line graph, including numbers, labels, and a title.

Discussion

1. What is the relationship between the length of a pendulum and its frequency?
2. Which has a higher frequency, the pendulum on a grandfather clock or the one on a cuckoo clock? Why?
3. How would you use your graph to predict the frequencies of other pendulums?
4. How does the line graph resemble the real graph made in *Swinging Bears, Part I?*
5. What other things have you learned from doing this activity?

Extensions

1. Explore the history of clocks.
2. Read about Galileo and his study of pendulums.
3. Explore some other aspects of pendulums.

Swinging Bears

How does the length of a pendulum affect its frequency?

Length of Pendulum in centimeters	Frequency of Pendulum in cycles per minute

What I found out: _____

_____

_____

My conclusions: _____

_____

_____

_____

Swinging Bears

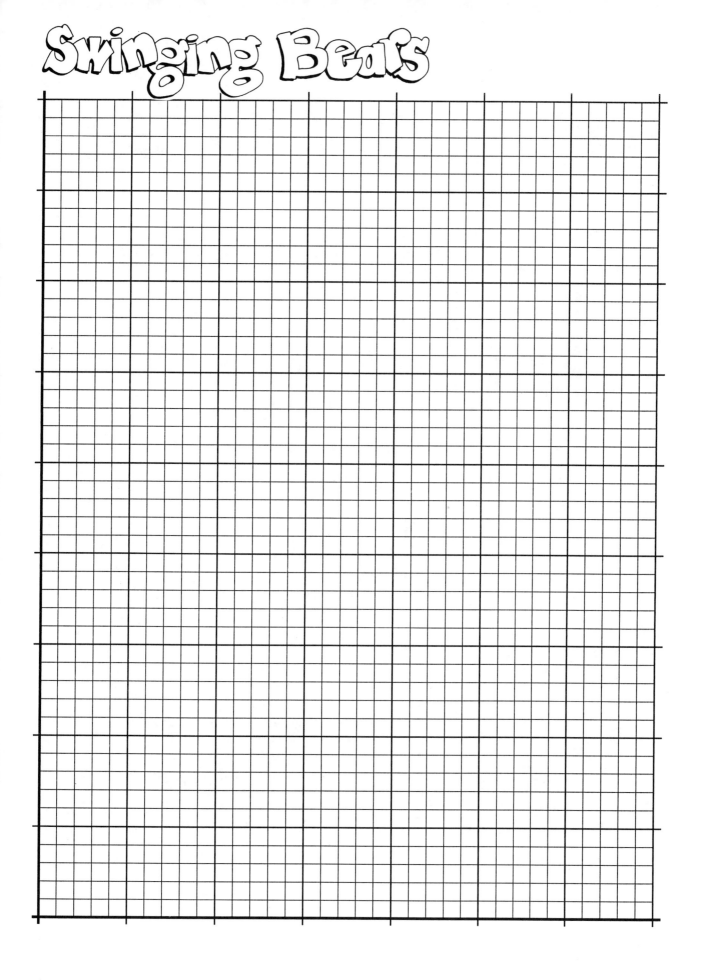

Testing the Bounce

The unifying thread which is woven through the following three activities is the testing of different variables which affect the bounce of a ball. Before doing any one or all three activities, students should spend time thinking about what might affect the bounce and how a fair test can be constructed.

One or more days before doing one of the ball activities, ask students, "What do you think affects the bounce of a ball?" Record their suggestions in a visible place (chalkboard, chart paper, etc.). From time to time during the day or week, maybe after a recess, raise the question again. Encourage students to discuss, debate, and do informal testing. Let their ideas simmer. Some of the variables it is hoped they will mention are height of drop, type of surface, kind of ball, how the ball is released, measurement accuracy, and temperature of the ball.

Once they are comfortable with their list, ask students to come to a consensus on which variable they would like to test first. The variable tested in each activity is as follows:

> **Have a Ball** - kinds of balls
>
> **On the Rebound** - starting heights
>
> **From the Ground Up** - types of surfaces

In addition, each activity focuses on another specific goal:

> **Have a Ball** - testing hypotheses
>
> **On the Rebound** - predicting based on data patterns
>
> **From the Ground Up** - "energy bank account"

"If we are testing _____ (name chosen variable), how can we make sure we have a fair test?" [A fair test involves controlling all but one variable. For example, to find how high different balls bounce as in *Have a Ball*, students must drop the ball from the same height and onto the same surface each time. If two variables are left uncontrolled, you cannot determine which is affecting the results—and to what extent.]

Have a Ball

See *Testing the Bounce* before doing this activity.

Topic
Testing hypotheses

Key Question
What kind of ball will bounce highest?

Focus
Students will explore the physical characteristics of balls (circumference, mass, and composition) and determine which one most influences the height of the bounce.

Guiding Documents
NCTM Standards
- *Make and use measurements in problems and everyday situations*
- *Collect, organize, and describe data*

Project 2061 Benchmarks
- *Scientists' explanations about what happens in the world come partly from what they observe, partly from what they think. Sometimes scientists have different explanations for the same set of observations. That usually leads to their making more observations to resolve the differences.*
- *Results of scientific investigations are seldom exactly the same, but if the differences are large, it is important to try to figure out why. One reason for following directions carefully and for keeping records of one's work is to provide information on what might have caused the differences.*
- *In doing science, it is often helpful to work with a team and to share findings with others. All team members should reach their own individual conclusions, however, about what the findings mean.*

Math
Estimation
Measurement
   linear
   mass
Order

Science
Scientific inquiry

Integrated Processes
Testing hypotheses
Observing
Controlling variables
Collecting and recording data
Comparing and contrasting
Generalizing

Materials
For each group:
   5-6 different balls (see *Background Information*)
   meter stick or tape
   balance scale
   metric masses
   string, optional
   crayons or colored pencils

Background Information
Mechanical energy, both potential and kinetic, as well as the force of gravity come into play when a ball is dropped and bounces off of a surface. But the focus of this activity is not on energy or gravity but on scientific inquiry, on testing hypotheses. Which of a ball's attributes most influences the height of its bounce? Three attributes that students can readily identify are circumference, mass, and the material used to make the ball.

Knowledge is the basis for forming hypotheses; they are not uninformed guesses. Normally a person will investigate a particular topic and then, based on the data gathered, generate hypotheses. Here students form hypotheses at the outset. How is this possible? Because previous games of jacks, four-square, basketball, etc. have given them experience with how balls bounce. Some sample hypotheses might be:
- The greater the mass, the higher the bounce.
- The smaller the mass, the higher the bounce.
- The larger the ball, the lower its bounce.
- The larger the ball, the higher its bounce.
- Rubber balls will bounce higher than those made of other materials.
- Balls made of harder materials will bounce higher than those made of softer materials.

Students will test a hypothesis, either by designing their own data collection or by gathering the data identified on the second activity page. Once the data is collected and the balls are ordered accordingly, students can look for patterns. Do the results confirm their hypotheses? If not, which of the balls' attributes seems to make a difference in the bounce? Is there a pattern of increasing or decreasing circumference? Compare masses. Is there a relationship between height of bounce and mass? What appears to make the difference?

32

After drawing conclusions, students should continue to question the results and construct further testing, broadening the sample base. The larger the sample, the more confidence can be placed in the accuracy of the results.

While circumference and mass can play a role, the composition of the ball (the material from which it is made and how it is constructed) is the attribute that most influences the height of the bounce.

Test a variety of balls, from small rubber balls and hi-bounce balls to balls used for team sports. The composition of regulation balls, which may not be what you have in the classroom, are as follows: basketball (leather with pebble grain), baseball (cowhide over wool yarn, rubber, and cork), softball (cowhide or horsehide cover over a soft material called kapok or rubber and cork), soccer ball (leather or other approved material), volleyball (leather), tennis ball (felt fabric of Dacron, nylon, and wool over rubber), table tennis ball (celluloid), and golf ball (rubber or synthetic material with dimples over a liquid sac or rubber).

Management

1. Choose the activity sheet that best meets your class' needs. The first sheet guides students in designing their own investigation. The second sheet provides more structure.
2. Decide the surface to be used and the height of the drop, somewhere between 100 and 150 centimeters.
3. Collect a minimum of two sets of balls, with five or six balls of different circumferences, masses, and composition in each set. Individual balls can be passed from group to group or, if there are enough sets, each group can have their own.
4. Make the other materials available in a part of the room where groups can gather what they need. Or set up three centers, one each for measuring circumference, mass, and bounce. Each center should have sufficient materials for more than one group to use at the same time.
5. Group members' jobs might include ball dropper, bounce measurer, and recorder.
6. To test the bounce, tape the meter stick to the wall. Hold the ball so its bottom is even with the designated height and let it drop; do not throw or push. By standardizing the way the ball is handled, a variable is being controlled.
7. To measure the bounce, find the distance from the surface to the bottom of the ball at the height of its bounce.
8. In the absence of tape measures, have students measure circumferences with non-stretchy string. The string can then be measured.

9. Students should conduct several trials with each ball because it requires practice to read a measurement when an object is in motion. When they are getting fairly consistent readings, they are ready to record the result. Students should read the measurement at eye level.

The following is offered for those students ready for more independent investigations.

Open-ended: introduce the *Key Question* and have student groups plan which balls they will test, identify characteristics that might determine the height of the bounce, make a hypothesis, collect and record data, and show their results in some way.
Guided: pose the *Key Question* and give each group the first activity sheet to guide them in testing a hypothesis.

Procedure

1. Display the balls to be used in front of the class. Have students make observations by holding up two balls (such as a tennis ball and a soccer ball) and asking, "How are these balls different?" [circumference, mass, composition, etc.] Ask the *Key Question*, "What kind of ball will bounce highest?" (Students may mention specific balls. Guide them toward predicting how the attributes they mentioned might affect the bounce.) "Does the circumference have something to do with how high a ball bounces? Does the mass affect the bounce? How about the material from which it is made?"
2. Model how to write a hypothesis as a statement (see *Background Information*). Have the class generate several hypotheses and write them on chart paper or on an overhead projector.
3. Distribute the activity sheet and have groups write the hypothesis they choose to test, the surface to be used, and the height of the drop.
4. Have each group conduct the bounce test, recording the type of ball and the height of the bounce on another paper. The bounce should be read several times before recording a result (see *Management 8*).
5. Instruct the groups to order the results from highest bounce to lowest bounce. (If the group has their own set of balls, they can use the actual balls in the ordering.) Students should draw and label the balls, in order, on the activity sheet and record the height of the bounce underneath.
6. Have students measure circumference and mass and describe the ball's composition as best they can.
7. Ask students to check the data to see if their hypothesis was confirmed. If so, have them examine the data for additional conclusions. If not, ask them to look for patterns in the data that lead to a new hypothesis. They should write their conclusions. (You may request that they write about how each attribute did or did not influence the height of the bounce).

8. Oversee the comparison of group results (if identical sets of balls were used) by asking one group to put the balls in order from highest bounce to lowest bounce. Groups should compare their results with those in the display and discuss discrepancies. What new questions are raised?

Discussion

1. How are these balls alike? [spheres]...different? [circumference, mass, composition, color, markings, etc.]
2. How high do you think the highest bounce will be? (have students predict)
3. We measured circumference, mass, and height of bounce. Which was easiest? Which was most difficult? Explain.
4. How do you know this was a fair test? [The way the ball was released, the kind of surface, and the height of the drop were kept the same. This isolated one variable, the type of ball, to be examined. A well-designed and fair scientific study tests only one variable at a time.]
5. What patterns can you find in the data? Based on the results, what new hypotheses can be formed?
6. What feature most affected the height of the bounce? What further testing could be done to confirm this result? Why is more testing necessary? [The larger the sample, the more accurate the results.]
7. Which part of this activity did you enjoy most? What surprised you?
8. In what other ways could the bounce be tested? [trying a variety of surfaces or dropping the ball from different heights] Make a hypothesis and design a way to test it.

Extensions

Do the activities, *On the Rebound* and *From the Ground Up*.

Teacher Resources

Doherty, Paul. "That's the Way the Ball Bounces." *Exploratorium Quarterly*. Fall 1991. Exploratorium 3601 Lyon St., San Francisco, CA 94123-9835

Taylor, B.A.P., Poth, J.E., and Portman, D. J. *Teaching Science With Toys: Physics Activities for Grades K-9*. Terrific Science Press. Miami University Middletown, Middletown, OH. 1994. (See "Bounceability," pgs. 179-185.)

Have a Ball

What kind of ball
will bounce highest?

Write several hypotheses of your own and choose one to test.

- The more mass a ball has, the higher it will bounce.
-
-

Kind of surface:

Height of drop:

Collect five or six balls. Record the data needed to test your hypothesis.

Conclusion:

What new questions do you have?

Have A Ball

What kind of ball will bounce highest?

Hypothesis:

Kind of surface:

Height of drop:

Measure the bounces and record on another paper. Draw and label the balls, in order, from highest bounce to lowest bounce.

Ball						

Height of bounce (cm)					
Circumference (cm)					
Mass (g)					
Composition					

Conclusions:

What new questions do you have?

On the Rebound

See *Testing the Bounce* before doing this activity.

Topic
Patterns (height of drop/ball's bounce)

Key Question
How does the ball's bounce compare with the height of the drop?

Focus
Students will discover a pattern relating the height from which a ball is dropped to the height of its bounce.

Guiding Documents
NCTM Standards
- *Use patterns and relationships to analyze mathematical situations*
- *Make and use measurements in problems and everyday situations*

Project 2061 Benchmarks
- *Mathematics is the study of many kinds of patterns, including numbers and shapes and operations on them. Sometimes patterns are studied because they help to explain how the world works or how to solve practical problems, sometimes because they are interesting in themselves.*
- *Measurements are always likely to give slightly different numbers, even if what is being measured stays the same.*
- *Graphical display of numbers may make it possible to spot patterns that are not otherwise obvious, such as comparative size and trends.*

Math
Estimation
Measurement
  length
Graphs
  bar and line
Patterns

Science
Physical science
  force and motion

Integrated Processes
Observing
Predicting
Collecting and recording data
Comparing and contrasting
Generalizing

Materials
For each group:
  golf ball
  meter stick
  small pieces of paper

Background Information
Students intuitively know that the ball will drop to the ground. The force of gravity is pulling the ball toward the Earth. Students also intuitively know that the higher the drop, the higher the bounce; the lower the drop, the lower the bounce. The attention here is on the pattern formed from the data.

We want students to get excited about finding patterns. There is a relationship, a pattern between the height of the drop and the height of the bounce for a particular ball striking a particular surface. With the kind of data being gathered, a line graph is often used to show the results. However, a bar graph is more understandable for younger students. If they compare the differences in bounce heights on a bar graph, students should find they form fairly consistent increments. They can then use this incremental distance to predict the bounce height for a drop from 120 centimeters.

Measurement is never exact. A measurement can always be taken to another, more precise decimal place. Measuring a ball in motion is even more difficult. Students should realize that their measurements are approximate.

Management
1. Divide the class into groups of three.
2. To test the bounce, hold the meter stick vertically or tape it to a wall or pole. Hold the ball so its bottom is even with the designated height and let it drop; do not throw or push. By standardizing the way the ball is handled, a variable is being controlled.
3. To measure the bounce, find the distance from the surface to the bottom of the ball at the height of its bounce.
4. Use a concrete surface if possible.
5. Although the graph starts with zero and rises to 100, more accurate measurements are likely if students conduct the tests in reverse order, starting with the 100-centimeter drop.
6. Students should conduct several trials at each height because it requires practice to read a measurement when an object is in motion. When they are getting fairly consistent readings, they are ready to record the result. Students should read the measurement at eye level.

7. To measure the graph increments, use a small piece of paper to mark the *difference in bounce height* between 0 and the 20-centimeter drop height (A). Then move the paper up to mark the *difference in bounce height* between the 20-centimeter and 40-centimeter drop height (B). Continue until all increments have been compared. The marks that were made should be fairly close together. Have students find the middle of the range of marks and mark that distance on the graph to predict the unknown bounce (C).

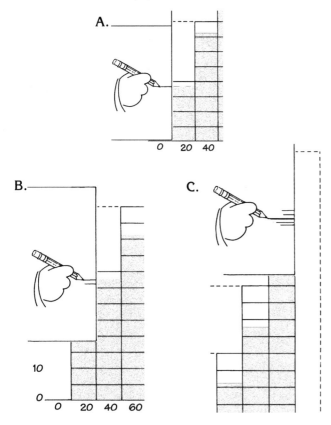

(The following is offered for those students ready for more independent investigations.)

> *Open-ended:* Ask the *Key Question* and have groups devise their own plan for answering it. Students should recognize the variables to be controlled–type of ball, kind of surface, and how the ball is released. They also need to choose an appropriate graph for the data.

Procedure

1. Give each group a golf ball and ask them to drop it (informally and with no measuring devices) different distances from the floor. Ask for their observations. "How do you think the height of the drop is related to the height of the bounce? Let's find out."
2. Distribute the activity page and instruct students to record the type of ball and type of surface (controlled variables).
3. Give each group a meter stick and have them go to the locations where they will be performing the investigation.

4. Direct students to record, in the first column, the drop heights indicated on the graph. Have them estimate, then test and measure the height of the bounce, continuing until they have completed the table.
5. Instruct students to make a bar graph from the data. Discuss their initial observations of the bar graph.
6. Ask, "Is there a pattern to how the graph is organized?" [counting by 20's] "Then what would be the drop height for the dotted bar?" [120 centimeters] Have students label the dotted bar.
7. Ask, "If the starting height is 120 centimeters, how can we predict the height of the bounce?" Give students time to think about this. Use one or more of their suggestions. Continue the discussion by asking, "If the height-of-drop bars increase consistently, do the bounce heights also increase consistently?"
8. Show students how to mark the distance between bounce heights (see *Management 6*) and find the middle of the range. Have them use their small paper to mark the predicted height of bounce for 120 centimeters. This is a prediction based on a pattern. They may want to fill in some of the cross bars at the top to help them determine the number of centimeters.
9. Instruct students to record data at the bottom of the activity sheet, then test and record the actual bounce. Hold a concluding discussion.

Discussion

1. What did you observe when you first dropped the balls (before doing the investigation)? [They fall to the ground (gravity). They bounce back up, but not as high. The higher you start, the higher the bounce.]
2. What does the bar graph tell us?
3. How do your group's results compare with others? (Variations in the accuracy of measurements and how well variables are controlled can cause differences.)
4. How can we make a height-of-bounce prediction for a 120-centimeter drop height? (Have students look for patterns in the bar graph.)
5. I wonder... [how the second bounce compares to the first bounce, if the second bounce also forms a pattern, if substituting another kind of ball will also make a pattern, how another type of surface might change the pattern, etc.] Design an investigation to find the answer to one of these wonderings.

Extension

This activity is one in a series that tests the different variables affecting a ball's bounce. See also *Have a Ball* and *From the Ground Up*.

Teacher Resource

Taylor, B.A.P., Poth, J.E., and Portman, D.J. *Teaching Science With Toys: Physics Activities for Grades K-9*. Terrific Science Press. Miami University Middletown, Middletown, OH. 1994. (See "Bouncing Balls," pgs. 73-77.)

On the Rebound

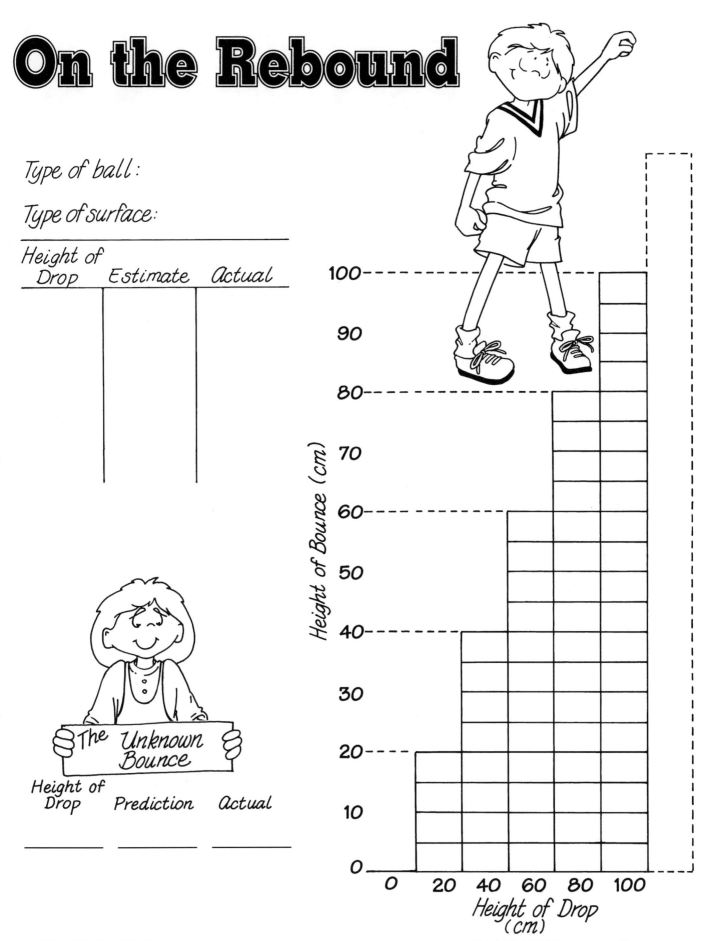

Type of ball:

Type of surface: _____

Height of Drop	Estimate	Actual

The Unknown Bounce

Height of Drop	Prediction	Actual

From the Ground Up

See *Testing the Bounce* before doing this activty.

Topic
Energy (kind of surface/height of ball's bounce)

Key Question
What effect do surfaces have on how high a ball will bounce?

Focus
Students will find that the kind of surface on which a ball is dropped affects how high it will bounce.

Guiding Documents
NCTM Standards
- *Collect, organize, and describe data*
- *Make and use measurements in problems and everyday situations*

Project 2061 Benchmarks
- *Scientists' explanations about what happens in the world come partly from what they observe, partly from what they think. Sometimes scientists have different explanations for the same set of observations. That usually leads to their making more observations to resolve the differences.*
- *Whenever the amount of energy in one place or form diminishes, the amount in other places or forms increases by the same amount.*
- *Offer reasons for their findings and consider reasons suggested by others.*

Math
Estimation
Measurement
  length
Order

Science
Physical science
  energy

Integrated Processes
Observing
Predicting
Collecting and recording data
Comparing and contrasting
Controlling variables
Interpreting data

Materials
For each group:
  golf ball
  meter stick

Background Information
As the ball is being held ready to drop, it has a given amount of energy. If no energy is lost after the ball is dropped, it will bounce back up to its starting height. So why does it not bounce that high? Some energy is taken away due to air resistance and some energy is absorbed by the surface on which it was dropped. The more energy that is transferred away from the ball, the lower the bounce. Since different surfaces absorb different amounts of energy, those which absorb the least energy allow the ball to bounce highest.

Let's look at an analogy. You have $100 and decide to go shopping at the mall. If you do not spend any of it, you still have $100 when you return home. Your balance remains the same. However, if you spend $70, do you have the same amount of money? [No.] Where did it go? [To pay for things you bought.] Your balance is now smaller.

Think of the ball as having an energy bank account. If no energy is "spent," the account stays the same. The more energy that is spent, the lower the account...and the bounce.

Management
1. Divide the class into groups of three. They can rotate through the locations with different surfaces.
2. To test the bounce, hold the meter stick vertically or tape it to a wall or pole. Hold the ball so its bottom is even with the designated height and let it drop; do not throw or push. By standardizing the way the ball is handled, a variable is being controlled.
3. To measure the bounce, find the distance from the surface to the bottom of the ball at the height of its bounce.
4. Students should conduct several trials on each surface because it requires practice to read a measurement when an object is in motion. When they are getting fairly consistent readings, they are ready to record the result. Students should read the measurement at eye level.
5. Choose surfaces such as concrete, carpet, dirt, blacktop, wood, floor tiles, foam pad, grass, metal plate, cardboard, etc.

Procedure

1. Have students brainstorm kinds of surfaces found around school. Pick five to use for the investigation.
2. Distribute the activity sheet and have students record the type of ball and height of drop to be used. They should write, in order from highest to lowest bounce, their predictions about which surfaces will allow the ball to bounce highest.
3. Have groups conduct the tests and record their data in the table.
4. Ask students to order the surfaces from highest to lowest bounce according to their data.
5. Discuss the results and introduce the idea of an energy bank account. Instruct them to write what they learned.

Discussion

1. How do you know this was a fair test? [The way the ball was released, the type of ball, and the height of the drop were kept the same. This isolated one variable, the kind of surface, to be examined. A well-designed and fair scientific study tests only one variable at a time.]
2. How does the bounce compare to the starting height? [The bounce is lower than the starting height.] Why does this happen? [Some of the ball's energy is taken away by air resistance and some is absorbed by the surface. The ball would need the same amount of energy to bounce back to its starting height but, since it has less, its bounce is lower.]
3. Does the kind of surface on which a ball bounces matter? Explain. [Yes. Different surfaces absorb different amounts of energy. The more energy they absorb, the lower the bounce.]
4. Do you think you could find a surface that would allow a ball to bounce as high as its starting height? Explain. [No. The ball would have to keep all the energy it started with and some energy is always changed or transferred when it strikes a surface.]
5. Would a ball ever bounce higher than its drop? Explain. [No, it would have to gain energy as it moved and this cannot happen.] What would happen if it could? [It would keep bouncing higher and higher and eventually go into outer space.]

Extensions

1. Try sand as a surface. What happens? [It hardly bounces at all.] Why? [The sand absorbs the ball's energy.]
2. Using the same surfaces, try a different ball (racquetball, a steel ball bearing, tennis ball, hi-bounce ball, etc.). A steel ball on a steel plate has a very high bounce.
3. Have students summarize the results of *Have a Ball, On the Rebound,* and *From the Ground Up* by choosing the combination of variables that will give the highest bounce (the best ball, the best surface, and the best height of those tried).

Teacher Resource

Taylor, B.A.P., Poth, J.E., and Portman, D. J. *Teaching Science With Toys: Physics Activities for Grades K-9.* Terrific Science Press. Miami University Middletown, Middletown, OH. 1994. (See "Bounceability," pgs. 179-185.)

From the Ground Up

Put the surfaces in order from highest bounce to lowest bounce.

Type of ball:

Height of drop:

Prediction	Actual

Surface	Height of Bounce (cm)

What did you learn?

Ball On a Roll

Topic
Kinetic energy
Inclined plane

Key Question
How does the ball's position on the ramp affect the distance it rolls?

Focus
Students will discover how slope affects the distance a ball will roll.

Guiding Documents
NCTM Standards
- *Make and use measurements in problems and everyday situations*
- *Construct, read, and interpret displays of data*

Project 2061 Benchmarks
- *The earth's gravity pulls any object toward it without touching it.*
- *Something that is moving may move steadily or change its direction. The greater the force is, the greater the change in motion will be. The more massive an object is, the less effect a given force will have.*

Math
Estimation
Measurement
  length
Averages
  median
Graphs

Science
Physical science
  mechanical energy

Integrated Processes
Observing
Controlling variables
Collecting and recording data
Comparing and contrasting
Generalizing

Materials
For each group:
golf ball
2 meter sticks (see *Management*)
several books
butcher paper
masking tape
piece of cardboard

Background Information
Dialoguing With Students: A Classroom Scenario

"What does the word *potential* mean?" [It means it could happen.] **Gravitational potential energy** is the energy an object has due to its position in a gravitational field. (Hold a ball on the ramp.) This ball sitting on the ramp has gravitational potential energy; it could move if I stopped holding it. The energy is not released until you let it go. The higher the ball is positioned on the ramp, the more potential energy it has.

As the ball rolls down the ramp, the potential energy is changed into kinetic energy. **Kinetic energy** is the energy an object has due to motion. "What do we know about moving things?" [•Moving cars can hurt you; they don't hurt when they are parked. •Big things are harder to move; football linemen are big because they do not want the other team to get passed them. •It is harder to catch a line drive than to pick up a baseball sitting on the ground.]

"Why run *around* a person on the playground?" [To avoid bumping into them and possibly hurting yourself.] "If you run into someone, what is the cause?" [You were moving.] You are different at rest than in motion. The faster you move, the more energy you have.

The greater the ball's energy, the farther the ball will travel once it leaves the ramp. A change in height causes a change in the distance a ball will roll. Let's look at what we mean by height. (Place the ball at 30 centimeters.) How far is it from the floor? (Now place the ball at 70 centimeters.) How far is it from the floor? Because the ball is higher off the floor at 70 centimeters, it has more potential energy.

Management
1. For the ramp, use meter sticks which do not flex or pieces of wood about one meter long. Tape the sticks at three or four places along the length, leaving a gap of about 2 cm between them. The golf ball will roll along this gap. If not using meter sticks, 10 cm increments need to be marked on the ramp.

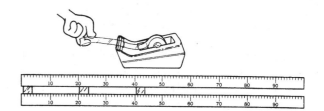

2. To set up the ramp, stack several books to a height of 10-12 centimeters. Place the 90-centimeter mark on the edge of the book stack, taped side at the bottom, and the other end on the floor. Roll out the butcher paper extending it to at least 6 meters, more if testing positions above 60 centimeters. Fold the edges of the paper up about two centimeters to keep the ball from straying.

3. To ensure a standard release of the ball, hold a small piece of cardboard at the designated point on the ramp with the ball behind it. Lift the cardboard and let the ball roll.

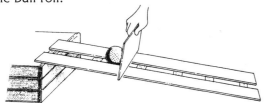

4. The leading edge of the ball is behind the starting line, so the distance it travels should be measured to the leading edge of the ball.

5. If you decide to set up just one or two ramps in the room and have groups rotate through them, you will need to devise a way for groups to distinguish their butcher paper marks from each other.

The following is offered for those students ready for more independent investigations.

Open-ended: Ask students the Key Question. Have them set up controls and carry out a plan to answer the question.

Procedure

1. Ask, "If I want the ball to roll a long way, does it matter where I put it on the ramp?" "How does the ball's position on the ramp affect the distance it rolls?"

2. Have students set up the ramp, show them how to release the ball (see *Management*), and have them make some practice rolls to determine which ramp positions they want to test. Reach consensus on the ramp positions everyone will use. For example, they might decide to try every ten centimeters from 20 to 60 (20-30-40-50-60).

4. Distribute the activity sheet and have students record the type of ball, the height of ramp where it intersects the books (for comparing group results), and the positions of the ball which will be tested.

5. Have students estimate and record how far the ball will roll from the lowest position on the ramp, then perform and record three trials. Students should continue until the table is completed.

6. Instruct students to cross out the shortest and longest distances in each set of trials and use the remaining numbers to complete the bar graph.

7. Lead a concluding discussion and have students respond to the question at the bottom of the activity page.

Discussion

1. Why do you think the ball rolled further the higher it was released? [It had more potential energy which was converted to kinetic energy when it moved.]

2. Place the ball in a position of low energy. (The ball should be moved to a lower part of the ramp.) How would you increase the kinetic energy in this ball? [Move it up the ramp.]

3. Compare the heights of the different ramp positions tested. (Students should measure.)

4. What makes the difference in results for the various ramp positions, the starting height or the distance the ball rolls on the ramp? [the height (Students may have a misconception that it is the distance.)]

5. Distance test: What would happen if we laid the two sticks on the floor and released the ball from different positions along the sticks? [There would be no difference. The ball would not move.]

6. Height test: What is your hypothesis if we released balls from the same height but different slopes? Design and perform this investigation. (When the slope becomes too steep, the ball may bounce when it gets to the bottom of the ramp. Some of its energy is taken away by the bounce. Slopes of 30° or less work best.)

7. How do your results compare with other groups?

Extension

Do the activity, "Rally Around the Room," found in the AIMS publication *Pieces and Patterns*.

Ball On a Roll

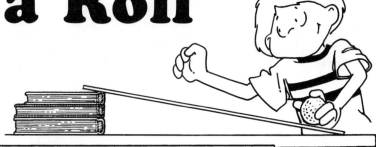

Type of ball :

Height of ramp :

Position of ball on ramp (cm)	Distance of roll (cm)			
	Estimate	Trial #1	Trial #2	Trial #3

Cross out the shortest and longest distances in each set of trials. Use the remaining number (median average) for the graph.

Position of Ball

Distance of Roll (cm)

What did you learn ?

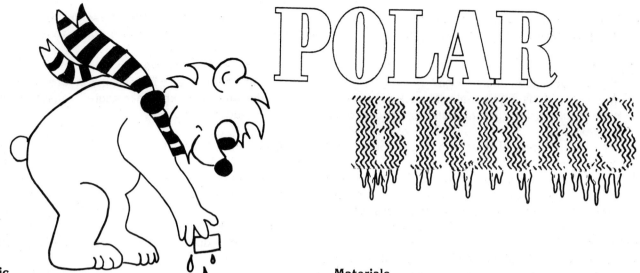

POLAR BRRRS

Topic
Insulation of ice

Key Question
How long can you keep an ice cube from melting?

Focus
Students will design ways to prevent an ice cube from melting.

Guiding Documents
NCTM Standards
- *Make and use measurements in problems and everyday situations*
- *Construct, read, and interpret displays of data*

Project 2061 Benchmarks
- *People can often learn about things around them by just observing those things carefully, but sometimes they can learn more by doing something to the things and noting what happens.*
- *Some materials conduct heat much better than others. Poor conductors can reduce heat loss.*
- *Heating and cooling cause changes in the properties of materials. Many kinds of changes occur faster under hotter conditions.*

Math
Measurement
    mass
Whole number operations
Graphs

Science
Physical science
    insulators

Integrated Processes
Observing
Collecting and recording data
Applying

Materials
For each group:
    1 ice cube (same size) in a plastic bag
    balance scale
    gram masses

Background Information
Pure water forms ice at 0° Celsius (32° Fahrenheit). When exposed to temperatures above 0°C (32°F), ice remains at 0°C, but it begins to melt. When ice melts, it absorbs heat energy from its surroundings. The challenge in this activity is to find an insulation material that will retard the absorption of heat energy.

Management
1. Initial planning by students should take place several days before the activity occurs.
2. Allow students who forget their materials to use available classroom supplies.
3. Allow one to two hours to elapse between the time the *Polar Brrrs* are set up and the time measurements are taken.
4. Groups of two or three are suggested.
5. Make ice cubes of the same size the day before.

Procedure
1. Have groups record information about their *Polar Brrrs* on the activity sheet.
2. Give each group one ice cube in a plastic bag. Have them find the beginning mass, put the cube in their *Polar Brrr*, and place the *Polar Brrr* on their desk or in a designated area.
3. At the given time, students should record how much time has elapsed.
4. Drain the melted water from the plastic bags.
5. Have students find the ending mass and write their own problem to show the amount of change in mass.
6. Facilitate the gathering of ending mass data from each group.
7. Groups should graph the data.

    46

8. Hold a class discussion and have students complete the questions on the graphing page.

Discussion
1. How did your ice cube change?
2. What kind of insulating material worked best? (The class could order the materials from most to least successful.)
3. What other kinds of materials might you try if the activity were repeated?
4. If you were taking refrigerated or frozen food to a picnic, how would you insulate it? Is your idea practical? (consider expense, bulkiness, etc.)

Extensions
1. Put ice cubes in different places to test melt rates (in the sun, in the shade, in the refrigerator, in a closet, etc.).

2. Have students find the mass of the ice cube *with* the melted water and compare that to the beginning mass.
3. Discuss how ice was obtained and kept before refrigeration. [Lake ice was cut in large blocks during the winter and kept in insulated houses covered with sawdust until used.]

Curriculum Correlation
Language Arts
  Compose a poem, make a cartoon strip, or write a story about the life of an ice cube.

Home Link
  Ask grandparents how food was kept cold when they were young.

NAME _____

POLAR BRRRS

HOW LONG CAN YOU KEEP AN ICE CUBE FROM MELTING?

TEAM MEMBERS _____
_____

ICE CUBE # _____

RULES: 1. YOU MAY BRING IN ANYTHING YOU WANT EXCEPT ELECTRICAL APPLIANCES, A THERMOS, OR AN ICE CHEST.
2. IT MUST FIT ON THE TOP OF YOUR DESK.
3. THE ICE CUBE MUST NOT MELT ALL OVER THE PLACE.

LIST THE MATERIALS YOU HAVE USED

_____
_____
_____
_____
_____

DRAW A PICTURE OF YOUR "POLAR BRRR" AND LABEL THE PARTS.

RECORD THE MASS OF YOUR ICE CUBE.

Beginning Mass (grams)

Ending Mass (grams)

WRITE A PROBLEM SHOWING HOW THE MASS CHANGED.

NAME_____

RECORD HOW MUCH TIME HAS ELAPSED. _____ HOURS

ENDING MASS
OF ICE CUBE
IN GRAMS

(empty grid plotted against the x-axis)

1 2 3 4 5 6 7 8 9 10 11 12 13 14

"POLAR BRRR" NUMBER

1. WHICH "POLAR BRRR" WORKED BEST? _____

2. WHY DID IT WORK BEST? _____

3. WHY DID YOU THINK YOUR "POLAR BRRR" WOULD WORK? _____
_____

4. IF YOU HAD A CHANCE TO BUILD ANOTHER "POLAR BRRR," HOW
   WOULD YOU DO IT DIFFERENTLY?
_____

_____

5. QUICK! REVERSE YOUR THINKING. WHAT IS THE QUICKEST WAY
   YOU CAN THINK OF TO MELT AN ICE CUBE? (FIRE IS NOT
   ALLOWED).

_____

_____

Cartons 'n Cotton

I. Topic Area
Insulation—Energy Conservation

II. Introductory Statement
Students will discover the effectiveness of insulation.

III. Math Skills
a. Measuring
b. Computing—Subtraction with regrouping

Science Processes
a. Gathering and recording data
b. Observing and classifying
c. Predicting and hypothesizing
d. Interpreting data

IV. Materials
(per group)
3 small jars with lids—all same size (large baby food jars work great)
3 half-gallon milk cartons
glue
cotton balls (about 250-300)
thermometer
hot tap water
worksheet

V. Key Question
How do we use a blanket or covering to keep things warm?

VI. Background Information
It is helpful for the teacher to know that the carton with the cotton on the inside will be noticeably warmer than the other 2 cartons.

VII. Management
1. Three class periods of 45 minutes each. It is better to make the insulated milk cartons one day and do the experiment the next. The math paper was completed the third day.
2. Groups of 4-6 are recommended. Size of groups should be determined by the number of thermometers and supplies available.
3. Before passing milk cartons out to the students, the teacher needs to cut a door large enough for easy access to the jars.
4. It is better to have three thermometers per group, but it can be done with just one.

VIII. Procedure
Day One
Assign groups. Pass out glue, cotton balls, and milk cartons. Students will glue cotton balls on the inside of one carton and on the outside of the second

carton. Be sure students include all sides, top, and bottom. The third carton will remain untouched.
Day Two
Collect all necessary materials. Give each student a worksheet. Go through "What the Students Will Do" step by step. As the students are waiting during the first 15 minute timing period have them sequence the steps gone through so far. The teacher can write these on the board for the students to copy. This gives the student a set of directions to use at home. Don't forget to include the gluing of cotton balls from the previous day. Record temperatures after the second 15 minute period. Discuss what is happening.
Day 3
Do computation on worksheet. Discuss results.

IX. What the Students Will Do
1. Students will insulate one carton by gluing cotton balls to the inside of the carton on all sides, top, and bottom.
2. Students will insulate one carton by gluing cotton balls to the outside of the carton on all sides, top, and bottom.
3. Students will leave the third carton untouched.
4. Fill all three jars with the same amount of hot tap water.
5. Put a thermometer in each jar and record temperature on worksheet. If group has only one thermometer work quickly but give the thermometer time to register in each jar.
6. Remove thermometer and place lids on jars.
7. Put each jar in a milk carton and close door.
8. Wait 15 minutes.
9. Remove jars and record temperatures one by one being careful not to mix the jars up.
10. Replace lids and return jars to same milk cartons and close the doors.
11. Wait 15 minutes.
12. Remove jars and lids. Record temperatures.
13. Discussion.
14. Do computation on worksheet. Discussion.

X. **Discussion**
   1. What is happening to the temperature of the water in each jar?
   2. Which jar is losing the most heat?
   3. Which jar retained the most heat?

XI. **Extensions**
   1. How do we insulate our bodies?
   2. Why do people in Alaska and the Arctic wear fur against their body? Why would they wear fur on the outside?
   3. How do we insulate our homes?
   4. How do we insulate pipes?
   5. Why does an animal's fur get thicker in the winter?
   6. How are animals insulated?

XII. **Curriculum Coordinates**
   Language Arts
   1. Sequencing and writing sentences about results.
   Art
   1. Design a poster asking people to insulate to save energy.

NAME_____

Cartons 'n Cotton

	BEGINNING TEMP.	15 MINUTE TEMP.	30 MINUTE TEMP.

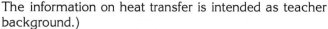

TINTS and TEMPS

Topic
Car color and temperature

Key Question
How do the temperatures of light and dark cars compare?

Focus
Students will discover that dark cars radiate more heat than light cars and that the temperature inside a closed car can rise to unsafe levels on hot days.

Guiding Documents
NCTM Standards
- *Make and use measurements in problems and everyday situations*
- *Collect, organize, and describe data*
- *Construct, read, and interpret displays of data*

Project 2061 Benchmarks
- *When warmer things are put with cooler ones, the warm ones lose heat and the cool ones gain it until they are all at the same temperature. A warmer object can warm a cooler one by contact or at a distance.*
- *Some materials conduct heat much better than others. Poor conductors can reduce heat loss.*

Math
Measurement
    temperature
Order
Graphs

Science
Physical science
    heat energy

Integrated Processes
Observing
Predicting
Collecting and recording data
Comparing and contrasting
Controlling variables
Generalizing
Applying

Materials
6-11 thermometers
1 meter stick
Cars (see *Management 2*)
Crayons or colored pencils

Background Information
(The italicized words indicate the key understandings which students should form while doing this activity.

The information on heat transfer is intended as teacher background.)

Heat flows from warmer things to cooler things until their temperatures equalize. The transfer of heat occurs in three different ways. In **conduction**, heat travels from a higher- to a lower-temperature region of a solid (or is transferred to another solid) by direct contact. Metals are good conductors of heat. In **convection**, heat is transferred when a heated liquid or gas physically moves to another location. For example, warm air is pushed up by colder air. In **radiation**, light rays strike an object and are reflected or absorbed. Rays that are absorbed will cause the object's temperature to rise. Heat from the sun warms the Earth's surface by radiation.

In many instances, heat is being transferred in more than one way. The surface of a car gains or loses heat by conduction, convection, and radiation. Surfaces vary in their ability to absorb radiation. *Dark-colored, dull surfaces tend to absorb and radiate heat better than light-colored, shiny surfaces.* A good absorber reflects very little light and therefore appears black. Light, shiny surfaces reflect more of the radiant energy and remain cooler.

The interior of a closed car also changes temperature by means of conduction, convection, and radiation. The glass allows light rays to enter. The radiant energy from the sun is absorbed by the dashboard, upholstery, seat belt buckles, etc. and travels throughout the material by conduction. The air in both the sunny and shady parts of the car interacts and reaches the same temperature by the process of convection. From a safety standpoint, *people, pets, and other living things should not be left in a closed, parked car on hot days.* The heat can become intense enough to cause death.

Hewitt, Paul G. *Conceptual Physics*. HarperCollins. 1989.

Management
1. Plan to do this activity sometime in the afternoon, during the warmer part of the day.
2. Make arrangements to use specific cars parked in full sun, including a white and a black or very dark car. You will need access to the inside of the car. If necessary, assure owners that you will be the only one placing the thermometers on and in the cars.
3. For the results to be valid, the thermometers need to register the same number of degrees in the same environment. With eleven thermometers, the external and internal temperatures of five cars and the outside air temperature can be taken simultaneously. With six thermometers, take the surface temperatures of five cars and the outside air temperature at the same time. Then take the cars' interior temperatures. To allow the thermometers to stabilize, keep the thermometers in place for fifteen minutes before recording a reading.

4. To take outside air temperature, tape the thermometer to one end of a meter stick. Hold the meter stick upright in a shady area away from buildings or other structures.
5. After the fifteen-minute stabilization period, read a car's interior temperature as quickly as possible after opening the car door. An open door allows the trapped air to escape and may lower the reading.
6. To facilitate the recording of data outdoors, have students place their activity sheet over a book, clipboard, or other hard surface.

The following is offered for those students who are ready for more independent investigations.

> *Open-ended:* Ask the *Key Question* and have student groups design their own investigations to answer it. They should carry out their plans and report the results to the class.

Procedure
1. Ask the *Key Question* and distribute the first activity sheet.
2. Have students write their predictions.
3. Discuss the procedure, including behavioral expectations and respect for personal property.
4. Take the class to the parking lot with their activity sheets, a writing surface, pencils, and the thermometers. Set one thermometer on the hood of each pre-selected car and another inside the car (preferably on the upholstery) in full sun. The outside air temperature should be taken in a shady spot away from buildings.
5. During the fifteen-minute wait, have students record the car colors and upholstery information in the table. Then ask the following questions:
   • What do you predict will happen?
   • What experiences have you had while traveling in, sitting in, or leaning against a car in hot weather?
   • What stories have you heard on television or read in the newspaper about car temperatures during hot weather?
6. Have students take the temperature readings of the cars and the air and record in the table. Readings should be taken with eyes perpendicular to the thermometer rather than at an angle.
7. Return to the classroom and have students examine their surface temperature data and sequence the car colors from coolest to warmest. They should write the color words on the lines and color the corresponding boxes.
8. Distribute the graph page and discuss what increments to use. Have them complete the graphs using the data in the table.
9. Instruct students to study the graphs and write their observations about surface temperatures and inside temperatures.
10. Hold a concluding discussion.

Discussion
1. How does a car's color affect its surface temperature? [Darker colors absorb and give off more heat energy than lighter colors.]
2. How might this affect you? [how comfortable you feel, how much the air conditioner is used (which increases the cost of operating the vehicle)]
3. How do the surface temperatures compare with the outside air temperature? [The surface temperatures are much warmer.] Why? [One thermometer was only measuring the temperature of the air while the car thermometers were registering a combination of the heat energy absorbed by the car's surface and the direct rays of the sun.] Why compare air and surface temperature? [It gives you an idea of how hot a car may be for a given air temperature.]
4. What variables, besides color, might cause changes in the surface temperature? [the kind of material used in the body of the car, the parking surface, whether the car is parked in the shade or sun, etc.]
5. How did the interior temperatures compare? (There may not be a definitive pattern related to color because many variables could affect the results. However, students should realize how hot the trapped air can become and the safety issues which result.)
6. What variables, besides color, might affect the inside temperatures? [the amount of glass (windows, sun roof), kind of upholstery, size of interior, whether the thermometer is placed in a sunny or shady part of the interior, the color of the car's surface, etc.]
7. How does the temperature inside a car compare to outside air? Why do you think it is so different?
8. How might this affect people and pets in hot weather? What safety tips would you give to others? [Don't leave a pet, child, or any living thing parked in a closed car during hot weather. It could kill them.]
9. For what other things, besides cars, might color affect the temperature and your comfort level? [clothing, houses, etc.] (Students may want to explore further, either through research or by planning new investigations.)

Extensions
1. Have students conduct temperature tests on cars parked in the shade. (The air temperature should be the same in order to directly compare cars in the sun with those in the shade.)
2. Test different parts of the same car. Does it make a difference where you place the thermometer?

Home Link
Based on the results of their investigation, have students look at the vehicle(s) their families own and decide the relative temperatures the cars would be compared to black or white. (Example: Our car would be hotter than a white car but cooler than a black car.)

TINTS and TEMPS

Name _____

How do the temperatures
of light and dark cars compare?

Prediction:

Time _____

Air Temperature _____

Color of Car					
Surface Temperature					
Color and Type of Upholstery					
Inside Temperature					

Based on surface temperature,
order the car colors from coolest
to warmest.

☐ _____
☐ _____
☐ _____
☐ _____
☐ _____

TINTS and TEMPS

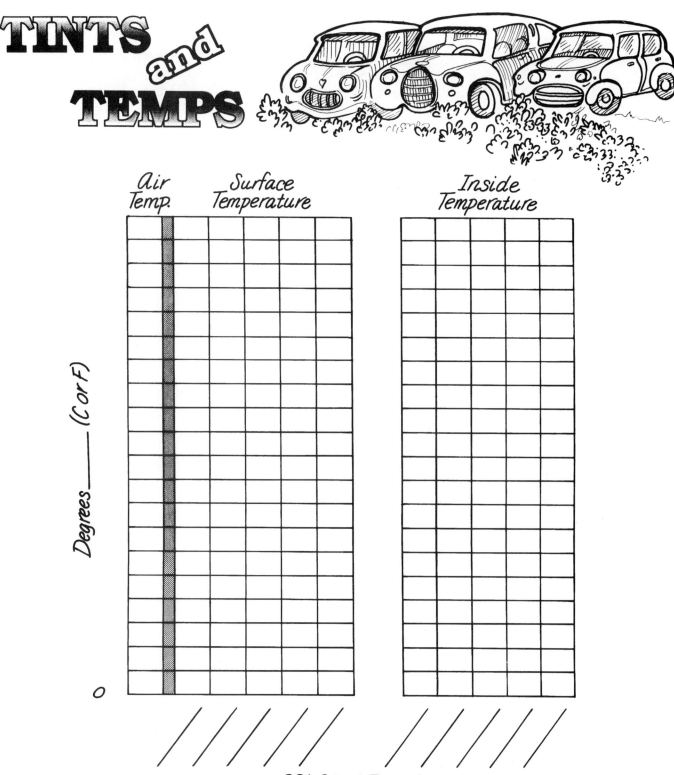

Air Temp. Surface Temperature   Inside Temperature

Degrees _____ (C or F)

0

//////////   //////////

COLOR OF CAR

Observations about surface temperatures:

Observations about inside temperatures:

Why Be A HOT HEAD?

I. Topic Area

Insulating the human head from the heat of the sun.

II. Introductory Statement

Students will learn which kinds of hats give them the best protection from the sun.

III. Math Skills

a. Subtracting
b. Graphing

Science Processes

a. Observing and classifying
b. Predicting and hypothesizing
c. Gathering and recording data
d. Interpreting data

IV. Materials

A varied assortment of summer hats (caps allowed, but hats preferred)
Student work sheets and class sheets for recording
A large class sheet to put on the wall
Pencils and crayons or colored pencils
Small Celsius thermometers

V. Key Question

Which hat or hats give the most protection from the sun's heat and why?

VI. Background Information

If the students haven't used a Celsius thermometer before, it would be helpful to show them how in Math class before doing this investigation.

The material the hat is made of, the color of the hat, the size and shape of it are important factors to discuss in evaluating the results of this investigation.

A child's own body heat can make a difference in the temperature reading. Everyone participating should be as cool and calm as possible.

VII. Management

1. This investigation will take at least 60 minutes. If more time is needed, the coordinating activities could be done in another class period.
2. This activity is done in and from the classroom. Groups of not more than 10 will go outside in turns to test their hats in the sun.
3. The thermometers will be placed on top of the heads and under the hats and they will be at room temperature so that should be recorded on the students work sheets.
4. Then when the students have spent 10 minutes in the sun, the new reading on their thermometer should be recorded on the work sheets and the first reading should be subtracted from the last reading to get the difference. It may vary from 0° to 20°.

VIII. Procedure

1. Ask the children to bring hats to school for a special "Hat Day."
2. Before that day arrives prepare a large class chart from manila graph paper.
3. List the children's names on it.
4. On "Hat Day" have the children model the hats so all may see.
5. Write the Key Question from section V. on the chalkboard and briefly discuss it.
6. Have the children make their predictions as to which hat will be the coolest to wear and mark the predictions on the large chart with a green felt pen.
7. Have them predict the warmest hat and mark it on the chart with a red pen.
8. Hand out the student work sheets and discuss them.
9. Next divide the children into groups of six to ten. The number of thermometers will dictate this.
10. Hand out the Celsius thermometers and have the students record the temperature on their students work sheets.
11. Put the thermometers on their heads and their hats over them.
12. Go outside to line up in the sunlight for 10 minutes.
13. After 10 minutes are up, hats come off and the thermometers should be read right away and the results recorded on the student work sheets.
14. Students should do the subtraction and record the difference on their sheets.
15. Have a class discussion after all have participated. See section X. Discussion.

IX. What the Students Will Do

1. Bring their own hats and model them.
2. Predict which hat will be the coolest to wear and which one will be the warmest.
3. Mark their predictions on the wall graph.
4. Record the room temperature from their Celsius thermometers onto the work sheets.
5. Put the thermometers on their heads and the hats over them.
6. Quietly stand in the bright sunlight for 10 minutes.
7. Quickly remove their hats and the thermometers when the 10 minutes are up.
8. Read the thermometers right away and record the temperatures on their work sheets.
9. Subtract their first temperature reading from the last one to find the difference.
10. Do the bar graph attached to the work sheet.
11. Take part in the class discussion.

X. **Discussion**

1. Which hat had the lowest temperature reading? The one with the smallest difference.
2. Which one had the highest reading? The largest difference? What type of hat? Whose?
3. Which type of hats make best protectors from the sun?
4. How could you maybe change your hat to make it better?

XI. **Extensions**

1. The students could change the colors of their hats by spray painting them. Then the investigation would mainly be based on which colors absorb the sun's rays and which ones reflect them.
2. Other modifications could be made on the hats by the students.
3. The children could make and design their own hats out of a variety of materials.
4. The students could do an experiment on the insulating differences of different colors and textures of human hair.

XII. **Coordinating Subjects**

Language

1. The students can write answers to a given set of questions about the investigation.
2. The students can write a few sentences of their own about the investigation.
3. The students can write their own paragraphs about it.
4. The students can write a creative story or poem about hats.
5. The students can do research reports about hats.

Art

1. The children can make hats out of paper and other materials. They could model them and vote on which one is the coolest, the most unique, the most useful, the most ridiculous, and the most well made.

Health

1. A lesson on human body temperature could involve the use of both Celsius and Fahrenheit thermometers.
2. A lesson on sunburns and sunstroke would be very timely during the hot season of the year.

Why Be A HOT HEAD ?

1. RECORD YOUR DATA BELOW.

2. SUBTRACT TO FIND THE DIFFERENCE.

TEMPERATURE AFTER THE 10 MINUTES	
TEMPERATURE BEFORE THE 10 MINUTES	
TEMPERATURE DIFFERENCE	

Why Be A Hot Head?

26
25
24
23
22
21
20
19
18
17
16
15
14
13
12
11
10
9
8
7
6
5
4
3
2
1
0

DEGREES CELSIUS

STUDENT NAMES

NAME

Curly Cue

Topic
Heat energy/air currents

Key Question
How does heat affect the air?

Focus
Students will discover that heat energy causes the air to move upward.

Guiding Documents
Project 2061 Benchmarks

"...*explore how heat spreads from one place to another and what can be done to contain it or shield things from it.*"(narrative for grades 3-5, pg. 83)

- *Things that give off light often also give off heat. Heat is produced by mechanical and electrical machines, and any time one thing rubs against something else.*

Science
Physical science
  heat energy

Integrated Processes
Observing
Comparing and contrasting
Interpreting data

Materials
2-3 different heat sources (see *Management*)
Thread
Transparent tape
Small paper plates, optional

Background Information
Heat moves. It spreads from one place to another in several ways. One way it moves is in air currents. When air is heated, as it is in this activity, it move upward. The effects of the upward-moving warm air can be observed on the paper coil. The more heat that is generated, the faster the air moves and the faster the coil will spin. Currents are evidence of convection at work. (When a heated substance itself moves, heat energy is being transferred by convection.) Both gases and liquids can have currents. See *Extensions* for other activities in which students can explore the concept of convection.

Management
1. Safety is always a consideration when heat is involved. Choose heat sources where there is no flame. Supervise students while they are using these objects.
2. Set up three areas in the room, one with a heat source turned off, one with a source generating low heat (such as a light bulb), and one where medium to high heat is generated (toaster, hot plate, etc.). Students can rotate through the three areas.
3. Students can cut the coil from their activity sheet or cut their own from a thin paper plate.

Procedure
1. Distribute the activity sheet and have them assemble the paper coil.
2. Review safety rules or ask students what safety rules should be established. Demonstrate how to hold the coil above the heat source.
3. Have each student hold their hand and then the coil in various positions around and above a heat source which has not been turned on. They should record the heat source, the amount of heat [none], and their observations [nothing happened, it felt like the rest of the air in the room].
4. Direct students to complete the table by repeating the process for the other two heat sources. The amount of heat should be recorded as none, low, medium, or high. Students should observe the relative speed of the moving coil and note the position in which the coil detected the warm air [above the source.]
5. Lead a class discussion and instruct students to write what they have learned.

Discussion
1. What do you think caused the coil to move? [warm air rising]
2. Where in relation to the _____ (toaster, etc.) did you notice air moving? [above it] How did the air feel? [warm]
3. How did your observations of the three heat sources compare? [The air did not move where there was no heat. The air moved upward over the ____ (toaster, hot plate, etc.) The lower the heat, the more slowly the air moved.]
4. If warm air rises, will the temperature be warmer near the ceiling than near the floor? Design an investigation to test this.
5. Where in your house would you want to sleep during the summer? (Mention possibilities such as upstairs or downstairs, top bunk or lower bunk.)

Extensions
1. To further help develop the concept of convection in both liquids and gases, do some of the following AIMS activities:
   "When Hot and Cold Meet" *Primarily Physics*
   "Side Talk" (sponge, helium balloon) *AIMS*, Voume X, Number 1
   a paper pinwheel in an air stream
2. Have students note that many objects such as light bulbs, toasters, and hot plates give off both light and heat. Ask them what other objects do the same.

Curly Cue

Cut and hang a thread
from the center point.

Object	Amount of heat	Observations

What did you learn? (Write on the back of this paper.)

PUFF MOBILES

Topic
Creative engineering/wind energy

Key Questions
1. How far can you blow your *Puff Mobile* in five seconds?
2. How can you improve your design for greater distance?

Focus
Students will construct a straw sail car (*Puff Mobile*) powered by their own wind energy. They will test and modify the car to achieve the maximum distance during five seconds of blowing.

Guiding Documents
NCTM Standards
- *Make and use estimates of measurements*
- *Make and use measurements in problems and everyday situations*

Project 2061 Benchmarks
- *Something that is moving may move steadily or change its direction. The greater the force is, the greater the change in motion will be. The more massive an object is, the less effect a given force will have.*
- *Moving air and water can be used to run machines.*
- *When parts are put together, they can do things that they couldn't do by themselves.*
- *Even in some very simple systems, it may not always be possible to predict accurately the result of changing some part or connection.*
- *There is no perfect design. Designs that are best in one respect (safety or ease of use, for example) may be inferior in other ways (cost or appearance). Usually some features must be sacrificed to get others. How such trade-offs are received depends upon which features are emphasized and which are down-played.*
- *Make something out of paper, cardboard, wood, plastic, metal, or existing objects that can actually be used to perform a task.*

Math
Estimation
Measurement
   length
Graphs

Science
Physical science
   energy

Technology
Engineering and design

Integrated Processes
Observing
Controlling variables
Collecting and recording data
Comparing and contrasting
Hypothesizing

Materials
For the class:
   stopwatch or timer with second hand

For each group:
   10 plastic drinking straws
   4 "wheels" (see *Management 1*)
   straight pins
   1 sheet of 8 1/2" x 11" paper
   several meter sticks or a meter tape

Background Information
This activity is an exercise in creative engineering that uses the wheel as a simple machine and a sheet of paper to "catch the energy" of the wind. Putting the *Puff Mobile* in motion demonstrates that wind can be used to perform work. Work is done whenever a force is used to move something.

The design and modification process, however, is the heart of this activity. Students begin with a challenge requiring divergent thinking: create a design which will move as far as possible in a linear direction. As they test and modify their *Puff Mobiles*, they make new hypotheses about what design changes will work. Through this testing and retesting, students should start to converge on a design that will best achieve this goal.

Management
1. For wheels you plan to reuse, purchase wooden beads with holes that will accomodate straws easily. If students are going to keep their *Puff Mobiles*, use mint-flavored Lifesavers® as wheels.

2. Copy the meter tape (in the back of this book) on bright paper in different colors. Assemble, laminate, and give two to each student. By linking the variously colored tapes together, students can easily see how many meters their *Puff Mobiles* traveled.

3. The day before, tell students to wear clothing that is suitable for working and racing on their knees.
4. Groups of two or three are suggested. Assigned jobs might include Puffer, Recorder, and Timer.
5. Time management
   a. Allow an extended amount of time to do the whole activity on one day.
   b. If you are going to organize the races as a tournament, construct and modify the designs one day and conduct the tournament on a second day.
6. If you wish to have more than one set of races, set up a tournament schedule so everyone has the opportunity to race at least twice. Make (or use the *AIMS Data Organizer/Binary Tree*) two sets of tournament brackets. After the first race, write the winners' names on one set of brackets and those who lost on another set of brackets (label *Consolation*, *Second Time Around*, or some other creative name). Continue racing until a winner for each of the two categories has been determined. Let them revel in their achievement; prizes are not necessary or advised.

7. Determine whether meters or centimeters will be used to record measurements. What rounding will be done?
8. The group's recorder might jot down problems observed during the trial. These can be referred to when recording and making design changes.
9. The graphs should be done after the three trials have been completed and the range can be determined. Though the graphs are separated, they must be numbered in the same way so comparisons can be made between them.

Procedure
1. Organize student groups and challenge them to build a car that, by blowing, will move the farthest.
2. Distribute the construction materials and the two activity sheets. Remind them that the paper and straws can be cut, but all components must be used.
3. Students should construct and name their *Puff Mobiles*. They should sketch their original design on the *Modified Mobiles* sheet.
4. Ask the first *Key Question*: How far can you blow your *Puff Mobile* in five seconds?
5. Have the class determine racing rules: starting position, from what part of the car to measure, number of puffers allowed, etc.
6. On the *Puff Mobiles* sheet, have students estimate and record the distance their vehicle will travel and decide who will be the Puffer.
7. Time their trials. Instruct them to measure the distance traveled and record it next to the estimate.
8. Ask the second *Key Question*: How can you improve your design for greater distance?
9. On the *Modified Mobiles* sheet, have the groups record at least one problem they want to correct, describe the change they will make, modify the *Puff Mobile*, and sketch the new design.
10. Students should estimate, test, measure distances, and make design changes again.
11. Have groups estimate, test, and measure distances one more time.
12. When all trials have been completed, have the class determine the range between the shortest and longest distances. They should decide how to number the graph so that all of the distances will fit. Each group's graphs must be numbered alike so comparisons can be made between their trials.
13. Instruct students to complete the three graphs on the *Modified Mobile* sheet, using their measurements on the first activity sheet.
14. Pair sets of *Puff Mobiles* and conduct the races. *Optional:* Organize and conduct a tournament (see *Management 6*).

Discussion
1. How important is the wheel? [very important to the development of modern civilization, as it has made transportation easier] How are the wooden beads like a wheel? [round]
2. Could your *Puff Mobile* have moved without wheels? [not as easily or as far] What might happen if the wheels were flat? [harder to move, greater friction]
3. Did a sail help? [Yes. It provided the surface against which the source of energy, human puffing, propelled the mobile.]
4. How did you change your *Puff Mobile* after the first trial? Why?
5. Did your *Puff Mobile* travel a longer distance after you made changes? Why or why not?

6. Besides design changes, what did you try that may have increased the distance traveled? [blowing technique, etc.]
7. Why do you think the winning *Puff Mobile* won? [elements of design, blowing technique]
8. What are the best parts of our designs that could be put together into one *Puff Mobile*?
9. In what other ways could we "puff" our mobiles? [a hair dryer, a fan]
10. What other kinds of vehicles use the wind to move?
11. What makes a real car move? [a gasoline engine]

Extensions

1. Pool the successful ideas from all of the groups and build an even better Puff Mobile. Challenge another class to a race.
2. Have each group report their best distance and make a class bar graph of the data, individually or on one large chart.
3. Try a 30-second race. A number of puffers will be necessary for each *Puff Mobile*.
4. Roll the *Puff Mobiles* down an incline plane. Record and graph results. How were the results different?
5. Design a *Puff Mobile* that will move on water. Agree on the materials to be used. (This could be used for assessment also.)

Curriculum Correlation

Literature

Ets, Marie H. *Gilberto and the Wind*. Puffin. New York. 1978. A small boy learns to play with and understand the moods of the wind.

Caney, Steven. *Steven Caney's Invention Book*. Workman Publishing. New York. 1985. Includes 35 stories of how people experimented and refined their designs, from earmuffs to roller skates.

Social Science

1. List as many uses for a wheel as can be found. [baby carriage, water wheel, pizza cutter, bicycle, grinding wheel, etc.]
2. Make a list of everything you see in one week that has wheels.
3. Research sail planes and other wind-powered vehicles.
4. Collect pictures of how we use wind as a source of energy.

Music

Sing *Puff, the Magic Dragon*.

PUFF MOBILES

Engineers _____
_____
_____

How far can you blow your Puff Mobile in five seconds?

_____
Name of PUFF MOBILE

Build a Puff Mobile using 10 straws, 4 "wheels", straight pins, and 1 sheet of paper. You may cut the straws and paper, but you must use <u>all</u> of the materials.

THE PRELIMINARY TRIALS

	Estimated Distance	Actual Distance
1.	_____	_____
2.	_____	_____
3.	_____	_____

FINISH

THE RACE

Estimated Best Distance

_____

Actual Distance

_____

66

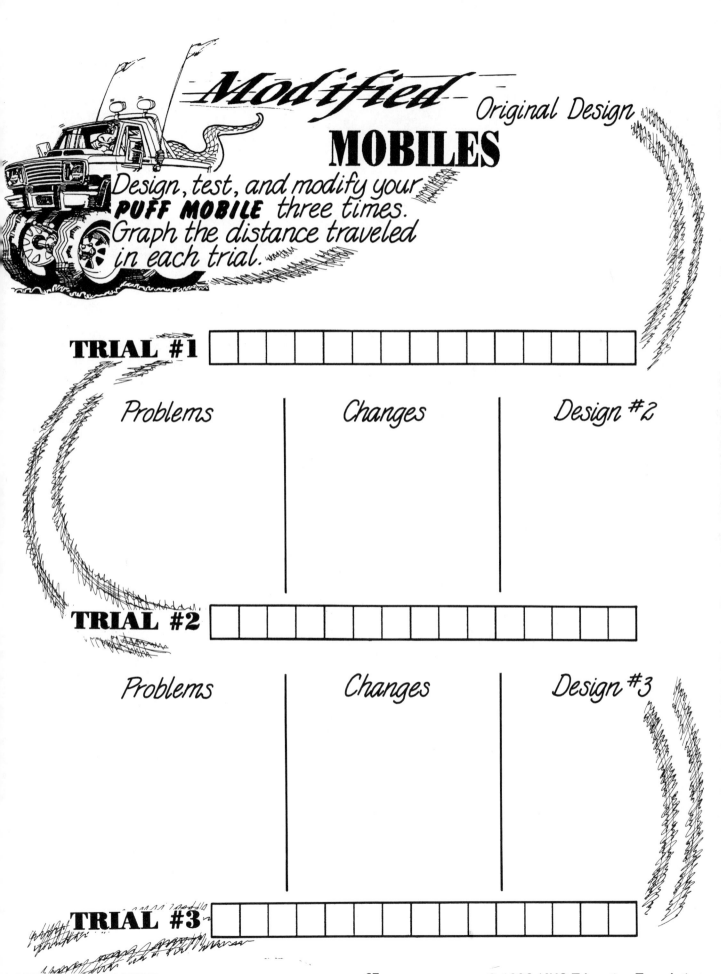

Modified Original Design

MOBILES

Design, test, and modify your **PUFF MOBILE** three times. Graph the distance traveled in each trial.

TRIAL #1

Problems | Changes | Design #2

TRIAL #2

Problems | Changes | Design #3

TRIAL #3

WIND Rollers

Topic
Kinetic energy of wind/wind patterns

Key Question
How does the wind make the roller move along?

Focus
Students will learn that the triangular points on the roller will act as sails that catch the wind and cause the roller to move.

Guiding Documents
NCTM Standard
- *Make and use measurements in problems and everyday situations*

Project 2061 Benchmarks
- *Air is a substance that surrounds us, takes up space, and whose movement we feel as wind.*
- *Something that is moving may move steadily or change its direction. The greater the force is, the greater the change in motion will be. The more massive an object is, the less effect a given force will have.*

Math
Estimation
Measurement
  length

Science
Physical science
  energy

Integrated Processes
Observing
Collecting and recording data
Comparing
Applying

Materials
Tagboard or 6" paper plates
Scissors
Metric tapes or sticks
Optional:
  metric trundle wheel

Background Information
Wind has kinetic energy, the energy of moving things. It comes from a Greek word for "move." The wind can push, pull, or lift things. The tagboard roller has potential energy or stored energy. When the wind pushes against the triangular sails or blades on the roller, it moves rapidly. The roller's potential energy has been changed into kinetic energy.

Several observations can be made about wind from watching the wind roller in action. There is a pattern of wind currents around buildings and on the playground. The speed of wind varies. Wind is not constant, but can die down and then start up again.

Management
1. This activity should done on a windy day; light breezes are not strong enough to make the rollers move.
2. Wind rollers can be made individually, but should be tested with a partner who can help with measuring.
3. The starting position for the wind roller does not matter. It can be set on the ground or thrown in the air.
4. Let the students set the rules for releasing the wind rollers and measuring the distance traveled.

Procedure
1. Tell students they will each be making a wind roller that will help them learn about the power and energy of the wind. Briefly discuss potential and kinetic energy.
2. Give students a tagboard copy of *Wind Roller Directions* and have them follow the directions to make their wind rollers. If they are using a paper plate, give them a regular copy of *Wind Roller Directions*; the diagram can be a guide for cutting the plate.
3. Distribute the activity sheets and have students estimate the distance their wind rollers will travel.
4. Group pairs of students and have them take their wind rollers, measuring instruments, activity sheets, and pencils outside.
5. Students should release their wind rollers three times, measuring and recording the distance traveled after each trial.
6. Have students evaluate their best trial and record their reasoning on the activity sheet.
7. Instruct students to gather and record distance data from three other classmates, compare this data to

their own, and think about making improvements in their wind roller.

8. Hold a concluding discussion.

Discussion

1. How did the wind make the roller move? [It pushes against the blades.]
2. What kind of energy does the wind have? [kinetic energy]
3. Before the roller began to move, what kind of energy did it have? [potential energy]
4. What wind patterns did you notice? [It blows at different speeds, it moves in different directions, etc.]
5. What else can the wind do besides causing our wind rollers to move? [breaks down rock, pumps water, generates electricity, etc.]
6. What variables might affect the distance traveled? [type of surface, amount of wind, rolling technique,

size of wind roller, material used to make the wind roller, etc.]

7. How could we improve our wind rollers?

Extensions

1. Form a team to make and test a better wind roller. Have a contest.
2. Build wind rollers of different sizes, with different numbers of blades, etc.
3. Build and test other wind catchers such as gliders.
4. Study about windmills or sailboats.

Curriculum Correlation

Language Arts

1. Write a story about some adventures you had while chasing your wind roller around the world.
2. Read selections from *Windy Day: Stories and Poems.* Caroline Feller Bauer, editor. HarperCollins. 1986.

Wind Roller Directions:

This pattern may be duplicated on tag board or used as a pattern with 6" paper plates.

1. Cut around the outside of the circle.

2. Punch a small hole in the center and cut each short line (radius) from the center to the edge of the inner circle.

3. Fold the points (sectors). Fold one up and one down alternately, until all are folded.

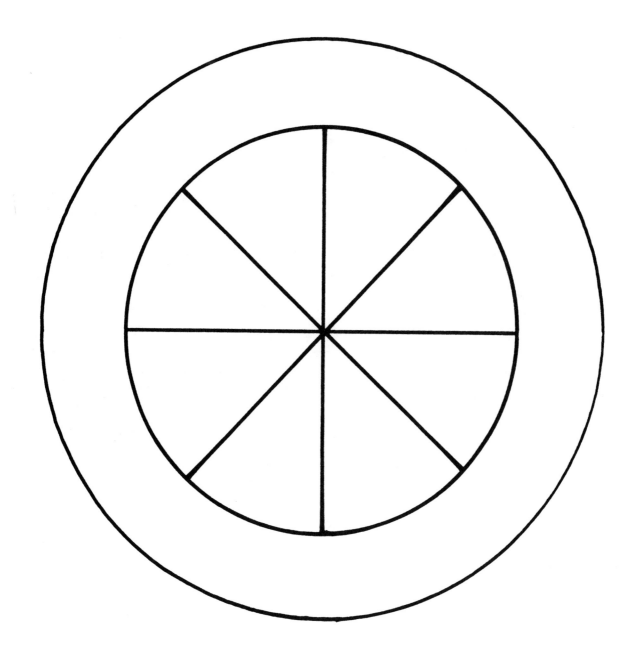

70

WIND Rollers

My wind roller loves a windy day.

*edict   My estimate_____ (how far I think it will go).

Measured distance for three trials:

*ther and
Record   1st trial _____   ✳ Put a star next to
          2nd trial _____   your best trial !
          3rd trial _____

*valuate   My best distance is _____. I think this is because

_____

_____.

*ompare   Collect "best distance" data from three other classmates
          and record.

_____
(name-distance)

_____
(name-distance)

_____
(name-distance)

WIND Rollers

Describe how these distances are the same or different from yours.

*
nalyze

_____

_____

_____

What changes would you make if you were entering a "wind roller" in a "longest roll" contest?

*
aluate
nd Apply —————————————————————————————

_____

_____

_____

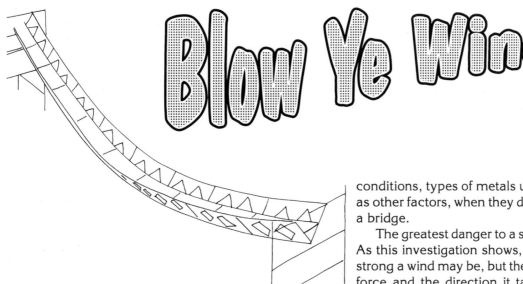

Blow Ye Winds

Topic Area
Physical Science—Wind force and suspension bridges

Introductory Statement
Students will learn about the effect of wind forces on suspension bridges and their stability.

Math Skills
Measuring
Estimating
Geometry

Science Processes
Observing and classifying
Controlling variables
Gathering and recording data
Generalizing and applying
Predicting

Materials
Ditto paper 81/2 " by 11 " (15 to 20 sheets)
Cellophane tape
Thread or thin string
Protractor
Scissors
2 meter sticks
Stick mounted on board
2 equal-size chairs with tall backs
2 hair blowers or fans with 2 speeds

Key Question
How does the force of wind affect a suspension bridge?

Background Information
Suspension bridges are supported by tall towers and suspended cables attached to a decking. These bridges are very graceful, yet they must support much weight. Engineers must be knowledgeable of weather conditions, types of metals used, wind forces, as well as other factors, when they design and construct such a bridge.

The greatest danger to a suspension bridge is wind. As this investigation shows, it is not necessarily how strong a wind may be, but the combination of the wind force and the direction it takes. At times a strong, steady wind can do less damage than lighter gusts coming from several directlons at once.

Modern suspension bridge design and construction must consider safety in all sorts of weather. The decking is designed to give stability to the entire suspension bridge structure. Box girders and trusses are used as stiffening supports for stability. Triangular shapes and girders are frequently used since the triangle is the strongest shape and cannot be twisted easily.

Modern decking and girders are often prefabricated, made in factories, and then hoisted into place above a river or bay. This method saves builders materials, money, and time. Modern decking is shaped more like the body of an aircraft, permitting wind to move freely around the deck and thus adding stability to the entire bridge structure.

Management
1. Suspension bridge of paper should be folded, cut, and constructed by the teacher or an adult, prior to the student investigation. Allow 30 minutes to do this.
2. Suspension bridge without triangular decking should be suspended carefully between two like-size chairs. Bridge should be suspended between the chairs by tying string or thread cables which have been inserted between folded down, taped sides of the bridge.
3. This investigation is best done as a total class activity. Allow 30 minutes for the testing of the bridge without and with the triangular decking at various wind speeds and from different angles.
4. Hair dryers or fans with two varying speeds will be used as the wind source. Fasten one metric ruler horizontally to the supported stick and place under the bridge center.
5. Several observers and recorders will be needed to watch and record the amount of sway at each speed and angle.

Procedure

1. Teacher will construct suspension bridge according to the attached directions prior to class investigation.
2. Teacher and student helpers will carefully suspend bridge between two identical-sized chair backs. The string "cables" will be tied to the chairs.
3. Teacher will guide discussion about the bridge construction, pointing out the v-shape and the fact that there is no decking.
4. Students and teacher will use the blower or fan at various speeds and angles to test the bridge stability. Results will be recorded on the chart.
5. Students will estimate the amount of movement or sway in cm with the wind source at various speeds when the triangular decking is added to make Bridge Two.
6. After the triangular decking pieces have been added to the bridge, students and teacher will use the blower or fan, the two wind speeds and various angles to test this bridge's stability. Results will be recorded on the chart.
7. Teacher will guide discussion, helping students determine the geometric shapes used in construction to stabilize the bridge and the various effects of the wind force.

What the Students Will Do

1. Students will examine the teacher made suspension bridge and participate in discussion about the first bridge design.
2. The first bridge, without decking support, will be tested for wind effect at the two speeds and angle changes. Two blowers can be used to produce the cross-wind simulation. Results will be recorded on the chart on the student page.
3. Students will predict what will happen to the bridge at various wind speeds and from various directions when the triangular decking supports are added. Estimates of sway in cm will be recorded.
4. Triangular decking pieces are added and students will test this bridge using the same blower speeds and angle directions. Results are to be recorded on the chart.
5. Students will compare data, predictions, discuss and make generalizations.

Discussion

1. What will happen to Bridge One when the wind blows at the different speeds and from different directions?
2. Why do you think this happened?
3. What do you think will happen when the triangular decking is added? Explain.
4. What were the results when you tested Bridge Two?
5. What happens when two wind sources are used from opposite directions?
6. Make some generalizations about wind force and its effect on a suspension bridge.

Extensions

1. Collect pictures and read about suspension bridges.
2. Invite a civil engineer or architect in to talk about design and construction.
3. Find out more about aerodynamics and wind.

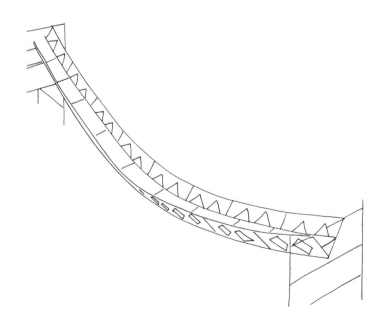

Blow Ye Winds

How does the force of wind affect a suspension bridge?

Speed	Angle of wind source	Bridge One (no decking)	Bridge Two (decking)	
			Estimates	Actual
High	Straight on →	cm	cm	cm
Low		cm	cm	cm
High	Left 45° ↘	cm	cm	cm
Low		cm	cm	cm
High	Right 45° ↗	cm	cm	cm
Low		cm	cm	cm
High	Above bridge	cm	cm	cm
Low		cm	cm	cm
High	Below bridge	cm	cm	cm
Low		cm	cm	cm
High	Two fans at opposite sides ↙ ↘	cm	cm	cm
Low		cm	cm	cm

What did you learn?

Which angle and speed most damaged the bridge?

Decking Pattern

Fold

Fold

Fold

Bridge Construction

1. Begin with a sheet of regular copy paper. Fold the paper in half from top to bottom.

21.5 cm

28 cm

2. Fold the paper in half from left to right.

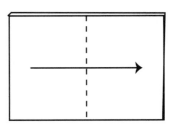

3. Fold the paper in half from right to left.

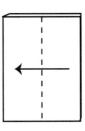

4. Your finished piece should be about 5.5 cm x 14 cm.

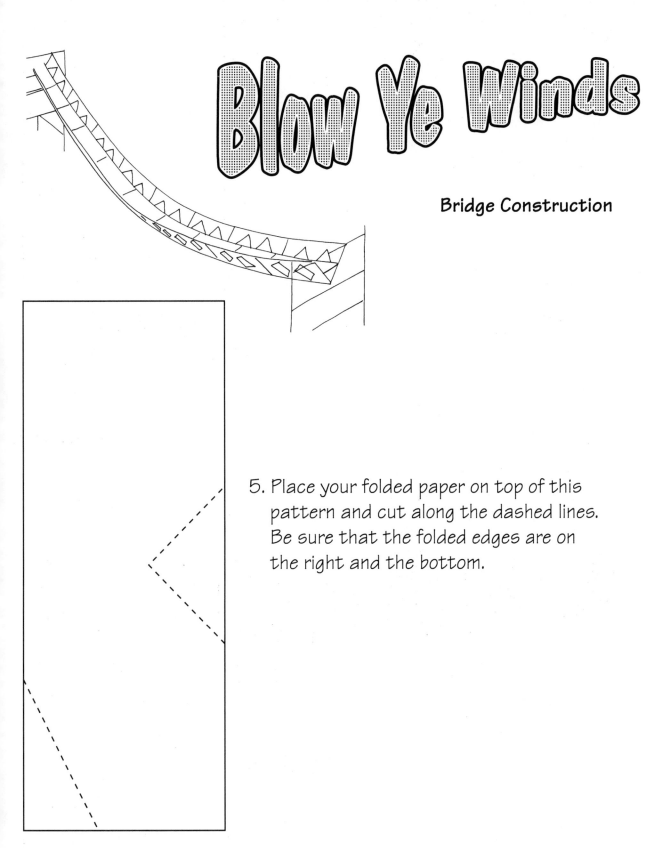

Blow Ye Winds

Bridge Construction

5. Place your folded paper on top of this
   pattern and cut along the dashed lines.
   Be sure that the folded edges are on
   the right and the bottom.

Make enough of these pieces to create a suspension bridge
of the desired length.

Blow Ye Winds

6. Tape several bridge sections together to make the desired length of bridge. Fold a small flap over along the top edge of the pieces.

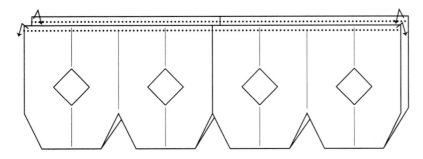

7. Place a piece of string under the flaps on each side and tape the flaps down.

String

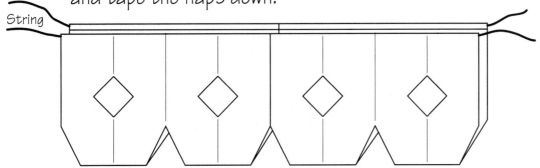

8. For Bridge Two, fold the decking pattern into a triangular prism and tape. Make enough decks to span the bridge.

9. Position the decks across the joints of the bridge.

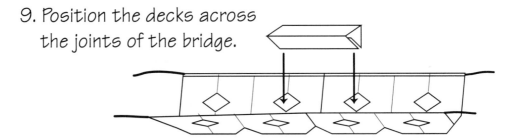

79

Meter Readers

Topic
Electric meters

Key Question
How much electricity does your family use?

Focus
This activity acquaints students with reading their home electric meters

Guiding Documents
NCTM Standards
- *Use mathematics in their daily lives*
- *Interpret the multiple uses of numbers encountered in the real world*

Project 2061 Benchmarks
- *People try to conserve energy in order to slow down the depletion of energy resources and/or to save money.*
- *Graphical display of numbers may make it possible to spot patterns that are not otherwise obvious, such as comparative size and trends.*

Math
Number sense and numeration
Whole number operations
    subtraction
Graphs
Estimation

Science
Physical science
    electricity
    energy conservation

Integrated Processes
Observing
Controlling variables
    time
Collecting and recording data
Comparing and contrasting
Generalizing
Applying

Materials
Electric meter at home
Calculators, optional

Background Information
    It was in the 1880's that electricity first became available for homes and work places—but only in large cities such as New York and London. It was mainly used for lighting. By the 1930's, electricity was more readily available. The first successful electric meter was made in 1888. A watt-hour meter, the basis for the kind used now, was patented in 1895.

    Electricity is brought from the source where it is generated through wires attached to tall poles. Wires branch off from these main wires and go to each house. They may be above ground or underground. At a house, the wires split into several different circuits.

    The electric meter records the amount of energy used by all the electrical circuits in a building. The energy is measured in kilowatt-hours, 1,000 watts of power for one hour of time. The formula for kilowatt-hours is **(watts x hours)/1,000**.

    Professional meter readers for the utility company go through about one month of training. The number of meters they read each day varies with the types of buildings and the distances between buildings. On residential routes in cities, 200-300 meters is an average load and includes five to seven miles of walking. Since apartment complexes group their meters, 1300 to 1400 meters can be read on one route. In rural areas, the route might include 150-250 meters and in the mountains, 50-75 meters.

    The newer digital meter is automatically read by a probe which is plugged into the meter. A code must be entered first. The utility company must have access to your property, granted by the installation contract, to take digital meter readings.

    In the near future, electric meters may be read directly from a central computer system using telephone lines.

    See *Reading the Electric Meter* for further information.

Management
1. Make a transparency of the fact sheet, *Reading the Electric Meter,* to better demonstrate how to read the dials.
2. Decide for what length of time the class will take readings. The activity sheet is designed to accommodate from one to seven 24-hour time periods.
3. You may wish to have students return their activity sheets after the first reading to clear up any questions and check for accuracy.

The following is offered for those students ready for more independent work.

Open-ended: Ask the *Key Question* and have small student groups plan what information they will record and graph for a specified period of time. Instructions for reading the meter will need to be given.

Guided Planning: Give each small group the *Meter Readers Plan Sheet* to help them organize their investigation of the *Key Question*. Students should be guided toward reading their home electrical meter to find out their daily usage. Instruction using the fact sheet, *Reading the Electric Meter*, will help answer the second question. The variable which should be controlled is the time the meter reading is taken each day.

Procedure
1. Ask the *Key Question*.
2. Distribute the fact sheet, read the directions, and have students practice reading the meters on the paper.
3. Give students the activity sheet to record their data. Have them record the date or day for the first reading on the table. Note that they will start on the last line and work their way up. This is to facilitate the subtraction and more clearly see the pattern of change.
4. Instruct students to fill in the time and the meter reading for the designated time period. Urge them to try to read the meter at approximately the same time each day.
5. On the first and last days, students should also draw the location of the dial pointers on the upper half of the activity sheet. This may be helpful to clear up possible errors in reading the meters.
6. At the end of the designated time period, have students return their activity sheets and find and record the amount of electricity used each day. They should also calculate and record the total kilowatt-hours of electricity used by subtracting the first day's reading from the last day's reading.
7. Give students the graph sheet and use the appropriate *Discussion* questions to guide them in setting up the bar graph for each days' reading. The days should be labeled along the horizontal axis and the kwh on the vertical axis. The range of kwh used will determine the increments. Each student's data will be different and it may or may not be possible to use common increments. Start numbering one or more increments below the first reading; otherwise, there will be no bar for the first day.

Example A: A first day reading of 21427 and a last day reading of 21482 would give a range of 55 kwh. Increments of three kwh allow all data to be plotted.

Example B: Increments of ten could be used for a range of 182 kwh.

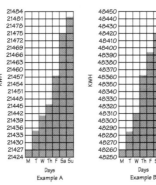

Days
Example A

Days
Example B

8. Students should complete their graphs.
9. Have students study their graphs and write statements about the information.
10. Complete the class discussion.

Discussion
1. Why are the pointers on the meter dials usually *between* two numbers rather than right on a number?
2. What labels do we need on the bar graph? [kwh and days of the week]
3. What shall we "count by" to label the kwh side of the graph? [It will depend on the range between their first and last readings and the number of lines on the graph.]
4. How can we find the cost of the electricity we used? [Call the local utility company or look at a utility bill to find the baseline cost per kwh. Then multiply by the number of kwh used.]
5. How much electricity does your family use?
6. What have you noticed about your graph pattern? [The bars go up, not down. The only exception is when the meter turns over every 1,000 or 10,000 kwh, depending on the kind of meter.] Why? [Each day you add on to the electricity used previously. You can't take back or subtract electricity already used.]
7. Were there any days where your use of electricity went up more than others? If so, why do you think so? [laundry day, baking day, etc.]
8. What difference in electrical use, if any, is there between families who live in apartments and those who live in houses? Why do you think there is a difference? [size of the place, a larger place takes more energy to heat or cool]
9. What else could affect the amount of electricity used? [energy efficiency of appliances, number of large appliances]
10. How does weather affect electrical use? [Heating in winter and air conditioning in summer use a major amount of energy. The more severe the weather, the more energy is needed to make people comfortable.]
11. In what ways are you already trying to save electricity at home?... at school? Name some more ways you can conserve electricity at your house... at school.

Curriculum Correlation
Language Arts
The following book traces the flow of electricity from a generator to the lamp in your home with easy-to-understand explanations and helpful illustrations. Its usefulness extends beyond the K-3 designation it is given.
Berger, Melvin. *Switch On, Switch Off.* Harper Collins. New York. 1989.

Art
Design a poster about a way to save electricity.

Home Link
Encourage students, together with their parents, to keep personal records of the information on their utility bill for several months. With parents' consent, the records, a graph, and conclusions could be turned in for extra credit.

Meter Readers
Plan Sheet

1. How will you find out your daily electrical use?

_____

2. What information do you need to carry out your plans?

_____

3. For how many days will information be gathered?

_____

4. What variables need to be controlled?

_____

5. Draw a simple sketch of the table you plan to use along with the table headings.

6. Draw a simple sketch of the graph (with labels) or other means by which you will show your results.

7. After seeing the results, what new questions do you have?

_____

_____

82

READING THE ELECTRIC METER

The electric meter measures the amount of energy used during a definite time period. The energy is measured in kilowatt-hours (kwh) which is 1,000 watt-hours. A kilowatt-hour is equal to ten 100 watt bulbs burning for one hour. The cost of a kilowatt-hour depends on the kind of fuel used to generate electricity and the cost of delivering it.

A meter records the energy used by all the electrical circuits in your home. Older homes may have four dials that turn back to zero after every 1,000 kwh used. They were installed during a time when fewer appliances were available and smaller amounts of electricity were needed. Newer homes have five dials that turn back to zero after every 10,000 kwh used. More electricity is needed today to run air conditioners and an increasing number of appliances. The newest meters are digital and are available, in some areas, by request.

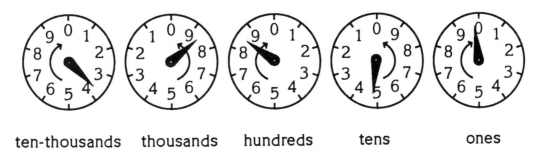

ten-thousands   thousands   hundreds   tens   ones

Notice that the numbers on the dials switch directions. Some dials have numbers reading clockwise and others counter-clockwise. When the pointer is between two numbers, write the lower number. When the pointer is between 0 and 9, read it as 9. The zero, in this case, stands for 10 and 9 is the lower number.

Professional meter readers from the utility company read the meter by the position of the pointers, rather than by looking at the numbers, much like you read a clock. They generally can read up to 20 feet away. Many meter readers also carry binoculars which extend their reading ability to 150 feet. A computer will reject their reading if it does not compare with past usage.

Meter Readers

Read your electric meter at the same time each day and record in the table. When a pointer is between two numbers, write the lower number. Draw the position of each dial pointer the first and last days you take readings.

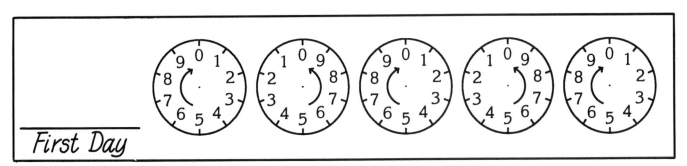

_____
First Day

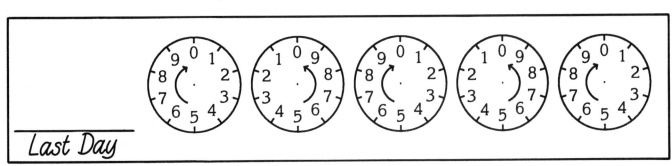

_____
Last Day

Total Difference _____ kwh

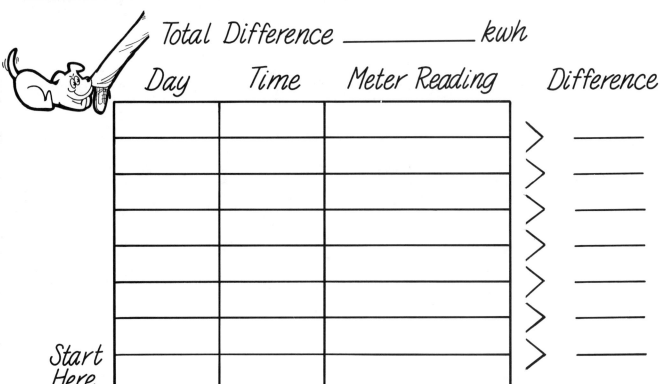

Day	Time	Meter Reading	Difference
			> _____
			> _____
			> _____
			> _____
			> _____
			> _____
			> _____
Start Here			> _____

Meter Readers

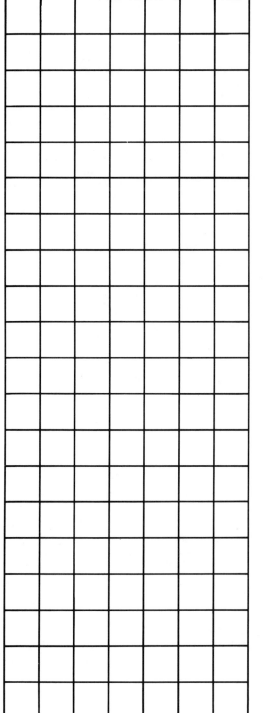

Design a bar graph to show your meter readings.

Write a statement about the number of kilowatt-hours used and number of days.

_____

_____

Write a statement about the cost of the electricity used.

_____

_____

Write one or more other statements about information on the graph.

_____

_____

_____

_____

Topic
Energy use-home appliances

Key Question
Which home appliances use the most electrical energy per hour?

Focus
Students investigate the watt variation of household appliances, a foundation for thinking about energy conservation.

Guiding Documents
NCTM Standards
- *Use mathematics in their daily lives*
- *Interpret the multiple uses of numbers encountered in the real world*

Project 2061 Benchmarks
- *Things that give off light often also give off heat. Heat is produced by mechanical and electrical machines, and any time one thing rubs against something else.*
- *People try to conserve energy in order to slow down the depletion of energy resources and/or save money.*

Math
Number sense and numeration
Whole number operations
Order

Science
Physical science
    electricity
    energy conservation

Integrated Processes
Observing
Predicting
Collecting and recording data
Comparing and contrasting
Applying

Materials
Appliances at home
Calculators, optional

Background Information
Watts are a unit of electrical power. Scientifically speaking, one watt is equal to one joule of energy per second or when one ampere of current is produced by one volt. A kilowatt-hour (kwh) is a unit of electrical energy equal to the power supplied by 1,000 watts for one hour. Kilowatt-hours are the unit of measure used by utility companies.

In the United States, regular household circuits are 120 volts. Appliances with high wattages, such as the clothes dryer and the range/oven, will be on a 240-volt circuit. The actual number of volts delivered on a circuit varies. For example, a 120-volt circuit may range between 110 and 125 volts.

Use the following table* as a general guide to the power rating of different appliances and the kilowatt-hours (kwh) used per year. These are *averages*. Individual use, geographic area, and the size and special features of an appliance offer the potential for a wide range of variation.

Appliance	Average Wattage	Average Hours Used Per Year	Kwh Per Year
Blender	390	40	15.6
Dishwasher	1,200	300	360
Range	12,200	100	1,220
Microwave	1,450	130	189
Toaster	1,200	35	42
Frostless refrigerator (12 cubic feet)	320	3,800	1,220
Frostless freezer (15 cubic feet)	440	4,000	1,760
Clothes dryer	4,800	200	960
Iron	1,000	140	140
Washing machine	500	200	100
Water heater	2,500	1,600	4,000
Electric blanket	180	830	149
Window fan	200	850	170
Hair dryer	750	50	37.5
Radio	70	1,200	84
Color TV, solid state	200	2,200	440
Clock	2	8,760	17.5
Vacuum cleaner	630	75	47
Light bulbs (on, in home)	660	1,500	1,000

* Gardner, Robert. *Energy Projects for Young Scientists.* Franklin Watts. New York. 1987.

The cost of operating home appliances can run from $500 to more than $1,000 a year. Energy prices for electricity, fuel oil, and gas have jumped in the last fifteen years. It is wise to use energy-efficient appliances. When buying a large appliance, the amount of energy the appliance will use should be considered as well as the sale price. *Energy Guide Labels*, which list the yearly cost per kwh, are required for all new refrigerators, freezers, water heaters, clothes washers, dishwashers, and room air conditioners. The expected lifetime of most large appliances varies from twelve to twenty years.

Compare the costs of owning two refrigerators for twenty years. *Refrigerator A can be bought for $650 and will use about $97 worth of energy per year. Its twenty-year cost will be $2,590. Refrigerator B costs $590 and will use about $112 worth of energy per year. Its twenty-year cost will be $2,830.* Although it will cost more to buy, Refrigerator A will save you money over its lifetime.

The biggest users of energy are the central heating system, water heater, and the refrigerator/freezer. But even the cost of using lights can add up to $50-$150 a year. Fluorescent bulbs are 3-4 times more efficient than incandescent bulbs and last 8-15 times longer.**

For further cost information, see the *Energy Efficiency* fact sheet.

** American Council for an Energy-Efficient Economy (Washington, D.C.) and Environmental Science Department, Massachusetts Audubon Society (Lincoln, MA). *Saving Energy and Money With Home Appliances*. 1985.

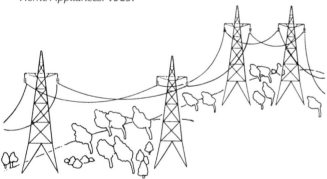

Management

1. You may change any of the appliances listed on the *Watts Going On* activity sheet if they are not typically found in the homes of your students by blocking out the appliance name and substituting another before making a classroom set.
2. Stress safety when searching for the wattages of appliances. Look for a sticker or plate on the outside of the appliance, on the edge of a door, or on an interior wall. If an appliance must be moved or lifted to gather data, **unplug it first**.

The following is offered for those students ready for more independent work.

> *Open-ended:* Ask the *Key Question* and have students plan their own investigation.
>
> *Guided Planning:* As students meet in planning groups, guide them with the following questions.
> 1. What appliances do you want to investigate?
> 2. What is your prediction? (Encourage a list of 1-3 appliances unless they are going to order the entire group of appliances.)
> 3. How are you going to find the electrical energy information? (This may not be answered until they start to search.)
> 4. What measure of energy will you use to compare the appliances? [probably watts]
> 5. What are group responsibilities?...individual responsibilities?
>   a) plan as a group but investigate individually
>   b) plan and investigate as a group, listing individual responsibilities
>   c) plan as a group, collect data individually, and average the results
> 6. How will you record the results?
> 7. After carrying out your plan, what conclusions and/ or questions do you have?
> 8. Using the information you have gathered, how can you better conserve electricity?

Procedure

1. Have students write their prediction about which appliances use the most watts per hour, ordering them from highest to lowest.
2. Give students the *Watts at Home* activity sheet to take home, complete, and return the next day. Emphasize that they are searching for the number of watts. If watts are not listed, have students look for amps and volts or kilowatts. Later this data will be converted into watts so the appliances can be compared.
3. The next day, have students transfer the data to the *Watts Going On* table. They should not attempt to fill in all of the columns, only those that match the information on the appliance label.
4. Students should then use the formulas at the top of the table, where necessary, to complete the watt column. Give guidance, if needed.
5. Have students use the number of watts to order the appliances again from highest to lowest and compare with their predictions.
6. Discuss the results and use the fact sheet to help students further understand energy consumption and possibilities for conservation.

Discussion

1. Which home appliances use the most electrical energy per hour? [probably the clothes dryer and range] Did everyone find this to be true?
2. Why do you think these appliances use more wattages? [The more watts, the more heat produced.]
3. What other home appliances, in addition to the ones we checked, might be big energy users? [central heater or furnace, air conditioner, and water heater]
4. Watts **and** hours of use determine which appliances use the most energy. Which appliances do you think use the most electricity during the year? [those with higher watt ratings that also run the most hours: central heater or furnace, central air conditioner, water heater, refrigerator/freezer, etc. According to one recent statistic, refrigerators alone use about 7% of the electricity in the United States.]
5. How does weather affect electrical use? [Heating in winter and air conditioning in summer use a major amount of energy. The more severe the weather, the greater the energy used.]
6. In what ways are you already trying to save electricity at home?... at school? Name some more ways you can conserve electricity at your house... at school. [The higher the watt rating, the more attention should be paid to controlling the use of the appliance. It also helps to buy efficient major appliances.]

Extensions

1. Convert watts into kilowatts (**watts/1,000**), and complete the kilowatt column in the *Watts Going On* table. Compute the cost of running each appliance for one hour, using local utility charges. Students will also experience one of the many ways decimals are used in the real-world.
2. Figure the cost of running an appliance for a certain number of hours. For example, how much would it cost to run the television five hours? Use **(watts x hours)/1,000** to figure kilowatt-hours. Then multiply by the local utility company charge for kwh.
3. Calculate the cost of using the appliances for one week, one month, or one year. First estimate or actually chart the number of hours the appliances are using electricity. Use **(watts x hours)/1,000** to figure kilowatt-hours. Then use your utility company base rate per kilowatt-hour to figure the costs. A table might have these headings:

Appliance	Watts	Hours of use	Kilowatt-hours	Cost per kwh	Total cost

4. Establish an electrical energy spending limit for one month (or a time period of your choice) and determine how long you can run each appliance and stay within your budget. How might it change how you live?
5. Take apart a non-working small appliance to see how it is constructed. A good reference: Macaulay, David. *The Way Things Work*. Houghton Mifflin. Boston. 1988.

Curriculum Correlation

Language Arts

Write or call the local utility company for kwh rates and more information on energy conservation.

Art

Design a poster about a way to save electricity.

Energy Efficiency *

Watts alone do not tell you which appliance uses the most energy. The number of hours of use must be considered.

BY THE HOUR

Cooling: window system	18-33¢
Cooling: central (3 ton)	55¢
Cooling: large fan	2-5¢
Portable heater	9-18¢
Oven: electric	16¢
Oven: gas	5¢
Rangetop burner: electric	15¢
Rangetop burner: gas	4¢
Microwave	15¢
Hair dryer	12¢
Iron	12¢
Vacuum cleaner	9¢
Toaster oven	6¢
Color television	2¢
100w incandescent bulb	1¢
27w compact fluorescent	.25¢
VCR	<1¢

BY THE MONTH

Electric central heater: smaller homes	$56-110
Electric central heater: larger homes	$114-400
Furnace: smaller homes	$16-40
Furnace: larger homes	$41-200
Electric water heater	$37-72
Gas water heater	$12-18
Freezer	$16-20
Refrigerator (16 cu. ft.)	$12-18
Refrigerator (20 cu. ft.)	$13-22
Waterbed heater	$9-22

BY THE LOAD

Clothes dryer: electric	61¢
Clothes dryer: gas	15¢
Electric blanket/a night	6-12¢
Dishwasher	6-11¢
Clothes washer	3¢

*Pacific Gas and Electric (PG & E), *Know Where Your Energy Is Going*. 1993. These are average figures based on rates of 12.3¢ per kwh (electricity) and 56.6¢ per therm (gas). Rates will vary across the country.

89 © 1996 AIMS Education Foundation

Watts
At Home

At home, find either watts, amps and volts, or kilowatts for these appliances. Record and label by each picture.

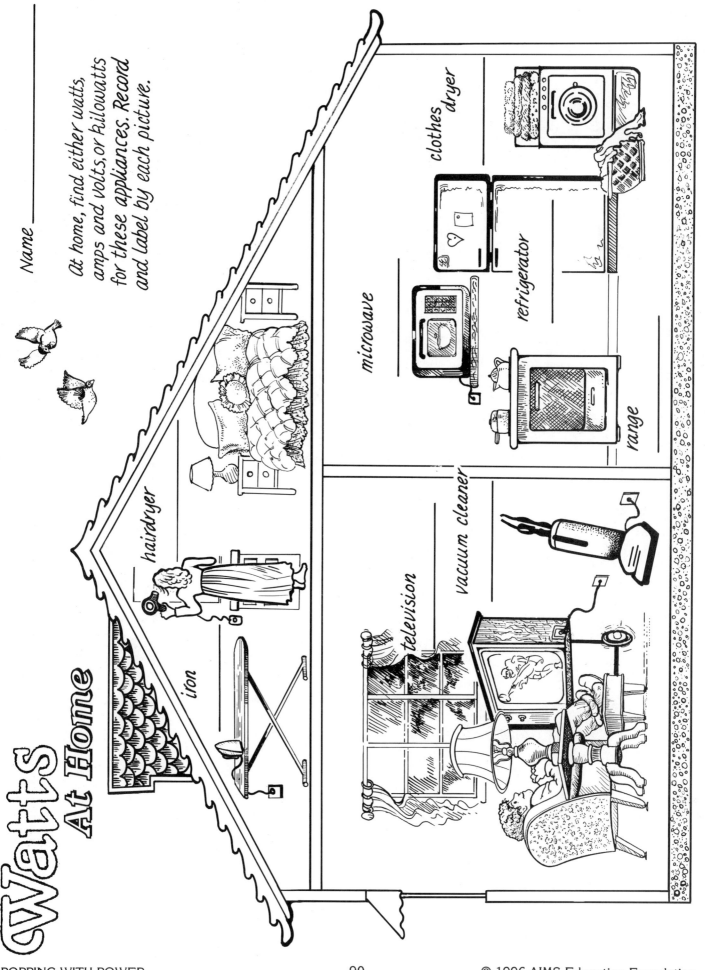

hairdryer

iron

microwave

clothes dryer

refrigerator

range

television

vacuum cleaner

90

Watts Going On

Name _____

1. Predict which appliances listed below use the most watts per hour. Put them in order from highest to lowest.

Prediction

1. _____
2. _____
3. _____
4. _____
5. _____
6. _____
7. _____
8. _____

Actual

1. _____
2. _____
3. _____
4. _____
5. _____
6. _____
7. _____
8. _____

2. Find the number of watts listed on these appliances and record in the table. If watts are not listed, record either amps and volts or kilowatts (kw). Use the formulas to find the number of watts, if needed.

Watts = Amps x Volts
Watts = Kilowatts x 1,000

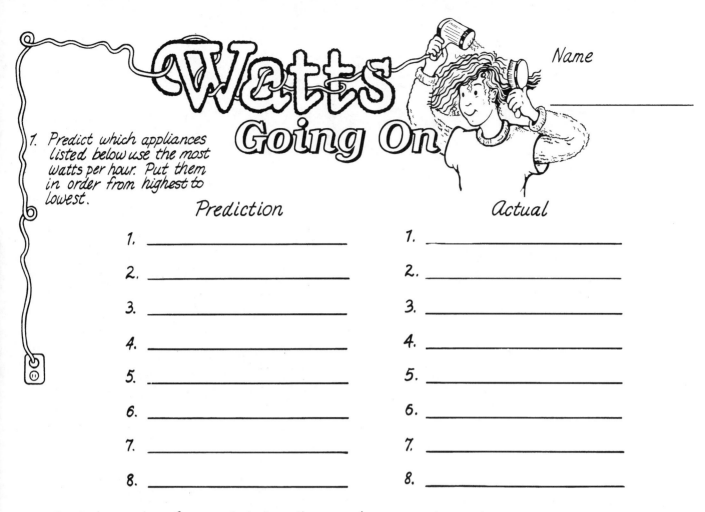

Appliance	Amps	Volts		KW		Watts
Clothes dryer						
Hair dryer						
Iron						
Microwave						
Range/oven						
Refrigerator						
Television						
Vacuum Cleaner						

3. Order the appliances again from highest to lowest, using actual results.

Topic
Energy use-light bulbs

Key Questions
1. How many light bulb watts are used in your house?
2. Do you live in a 3,000-watt house?

Focus
Students will take a home survey of light bulbs, then design an energy-saving lighting system for a home.

Guiding Documents
NCTM Standards
- *Use mathematics in their daily lives*
- *Interpret the multiple uses of numbers encountered in the real world*
- *Systematically collect, organize, and describe data*

Project 2061 Benchmarks
- *Things that give off light often also give off heat. Heat is produced by mechanical and electrical machines, and any time one thing rubs against something else.*
- *Some energy sources cost less than others and some cause less pollution than others.*
- *People try to conserve energy in order to slow down the depletion of energy resources and/or save money.*

Math
Estimation
Counting
Whole number operations
Decimals
Graphs

Science
Physical science
   electricity
   energy conservation

Integrated Processes
Observing
Collecting and recording data
Comparing and contrasting
Interpreting data
Generalizing
Applying

Materials
Light bulbs at home
Calculators, optional

Background Information
An incandescent bulb generates a great amount of heat to produce light. Only 5-8% of the energy produced is converted into light; the rest is dispersed as heat. The number of lumens determines the brightness of the light. The higher the number of lumens per watt, the higher the filament temperature and the whiter the light. The tungsten filament heats to temperatures of 2,000-2,500°C (3,600-4,500°F) turning it white-hot. Incandescent bulbs burn out because the filament weakens from being repeatedly heated to high temperatures and then cooled.

Fluorescent bulbs, although more expensive, make up the cost in energy savings. They can generate the same number of lumens as an incandescent bulb at a significantly reduced wattage. Some city ordinances require them in certain parts of new homes and utility companies are encouraging their use.

Halogen bulbs offer another alternative for saving energy. They are more efficient and last longer than incandescent bulbs; however, they are not as cost effective as fluorescent bulbs.

A Bit of Electric Light History
The arc light, in which sparks of electricity jump continuously between two carbon rods, was being used for street lights in the late 1800's. However, an arc light needed replacing every few hours, was blinding (4,000 candlepower compared to the 10-20 candlepower of gas lights), smelled badly, caused fires, made loud noise, and was expensive. Improved arc lights are used today where bright, concentrated light is needed: arc welding, searchlights, theater spotlights, etc.

The search was on for something different. Although Thomas Edison is given credit for inventing the electric light bulb in 1879, there were over twenty other people, particularly Englishman Joseph Swan, who achieved some measure of success with the idea. What set Edison apart is that he not only invented the bulb, but also a whole electrical system, including generators, meters, switches, and fuses, to bring it to the general public.

After testing thousands of filament materials such as corn, silk, metal, straw, human hair, and splinters of wood, Edison was able to keep an electric light burning for about 40 hours. He used a carbon-coated cotton thread as a filament and a glass bulb from which the air had been removed. He later used a filament made of bamboo fiber to increase the bulb life to 240 hours.

(See *Light Bulbs: The Inside Story* and *Which is Better?* for further information.)

Management

1. *Part One* involves individual students taking a home survey of light bulbs and examining the construction of one-way and three-way bulbs. In *Part Two*, teams design energy-saving lighting for a home. The more advanced *Part Three, Which is Better?*, adds a comparison of incandescent and fluorescent light bulbs.
2. Caution students to count only the light bulbs in sockets, preferably when they are turned off. Do not count bulbs being stored for future use.
3. You might suggest that students leave some space in their table for figuring totals.
4. The local rate for kilowatt-hours (kwh) needs to be obtained.
5. Consider making transparencies of selected activity sheets.
6. For *Part Two*, run the scenario cards on card stock and cut apart.
7. *Which is Better?* is a more challenging activity which requires several independent computations to complete the graph (see *Procedure*). To save time, you may wish to find the cost of incandescent and fluorescent bulbs.

The following is offered for those students ready for more independent work.

> *Open-ended:* Ask the *Key Question* and have students plan how they will collect, record, and graph the information. Then use *Light Bulbs: The Inside Story* and/or *Which is Better?* to enhance their learning.

Procedure
Part One
1. Ask the *Key Questions* and have students circle the estimate of total light bulb wattage on the activity sheet.
2. Ask "What kind of information do we need to record?" [location, number of bulbs and wattage] Have students build the *Light Bulbs at Home* table, complete the survey, and bring it back the next day.
3. Direct students to compute the total watts and describe the procedure they used on the activity sheet. [counting like-wattages and multiplying, figuring wattage in each room and adding, etc.]

4. Use these questions to guide students toward writing an equation:
   a. What information do we need to know to figure the cost of using our light bulbs? [our wattage, utility company unit of measure, rate of that unit of measure, number of hours of use]
   b. With what unit does the utility company measure energy usage? [kilowatt-hours (kwh)]
   c. How do we change watts into kilowatts? [watts/1,000 = kwh]
   d. How can we find out our rate per kilowatt-hour? [look on a bill; call the utility company; **(watts/1,000) x kwh rate = energy use for one hour**]
5. Use these questions to guide students in constructing the class graph, *Glowing Report*:
   a. "How should we label the axes of our graph?" [wattage ranges on the horizontal axis, number of homes on the vertical axis]
   b. "What increments shall we use?" [depends on the class range of total watts]
6. Students should complete the class graph and write statements based on the data.
7. Together with students, study *Light Bulbs: The Inside Story* and have them complete the three-way watts pattern and circuit coloring. (See *Discussion* for related questions.)

Part Two
1. Organize students into teams of three or four. Distribute the house sheet and one scenario card to each team.
2. Teams should then draw their lighting design, using the setting and restrictions set by the scenario card. They will need to determine total watts used when they complete the project.

Part Three (Which is Better?)
1. Together with students, read the information given. Have students find and record the cost of incandescent and fluorescent bulbs in your area.
2. To compare costs over time and construct the bar graph, students will need to do the following:
   a. Decide what incandescent and fluorescent bulb wattages will be compared. To be equivalent, the bulbs must have a similar number of lumens.
   b. Determine how much one incandescent bulb costs.
   c. Calculate the number of incandescent bulbs that will be used over 9,000 hours. Use the average life listed on the package.
   d. Calculate the energy cost. One example:

$$\text{(watts x hours)}/1,000 = \text{kwh}$$
$$\text{kwh x kwh rate} = \text{energy cost}$$

   e. Add the bulb and energy costs together. Use these totals to determine the increments for the bar graph.

f. Make a key and graph the cost of the bulbs. Graph the energy cost, continuing up the same bars. The top line of each bar should match the totals.

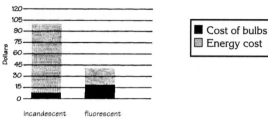

Discussion
1. What problems, if any, did you have getting the light bulb information? [how to get wattage from recessed light fixtures, how to deal with three-way bulbs when figuring total wattage, etc.] How shall we solve these problems?
2. Which light bulb wattage was most common at your house?...in all our houses?
3. How many light bulb watts are used in your house?
4. Where do you need brighter lights?...dimmer lights?
5. If all your light bulbs were on, how much would it cost per hour?
6. What statements can we make from the data on our graph?
7. In what ways could you save electricity?

Light Bulbs: The Inside Story
1. Using each of the possible positions, describe the brightness of a 50-100-150 watt bulb in which the *lower-wattage filament* is no longer working. [Position 1: no light; Position 2: 100 watts; Position 3: 100 watts]
2. Using each of the possible positions, describe the brightness of a 50-100-150 watt bulb in which the *higher-wattage filament* is no longer working. [Position 1: 50 watts; Position 2: no light; Position 3: 50 watts]

Which is Better?
1. What do you notice about the number of lumens in relation to watts? [As the number of lumens increases, so does the number of watts.]
2. What is the benefits/drawbacks to using incandescent bulbs?...fluorescent bulbs?
3. What conclusions have you reached? [Over time, fluorescent bulbs save more money and energy.]

Extensions
1. Figure the average total number of light bulbs or average total wattage based on class data.
2. Have students write their own scenarios with these instructions: Describe a house or other living space (number of bedrooms and bathrooms) and the people who live in it. Then design their home so that it uses no more than _____ watts in lighting.
3. Collect various used light bulbs, carefully break them open, and examine the circuit structure. On a three-way bulb, notice the two contact areas on the outside of the base. If students are to handle the bulbs, have them wear gloves.
4. Invite a utility company person to discuss the energy demands of incandescent versus fluorescent lights.
5. Repeat *Which is Better?* using bulbs of different wattages.

Curriculum Correlation
Language Arts
1. Write or call the local utility company to get the kwh rate and more information on energy conservation.
2. Do further research on incandescent, fluorescent, and halogen bulbs. There is an excellent article on this topic in the October 1992 issue of *Consumer Reports*.
3. Read about Edison's life in *The Story of Thomas Alva Edison* by Margaret Davidson and *Thomas Edison and Electricity* by Steve Parker.

Name _____

Do you live in a 3,000-watt house?

1. Circle the number of light bulb watts you estimate are used in your house.

   1-1,000   1,001-2,000   2,001-3,000   3,001-4,000   4,001-5,000

2. Collect and record information about all of your light bulbs in lamps and
   ceiling fixtures, including the room where you found them.

Light Bulbs At Home

3. Total watts_____
   How did you find your total?

4. Show a way to figure the cost of
   using all your light bulbs for one
   hour.

Light Bulbs: The Inside Story

A one-way incandescent light bulb has two places or contact points to which the wires inside the bulb are connected-the center of the base and the outer shell. Linking the two wires is a filament, the part that glows when electricity flows through it.

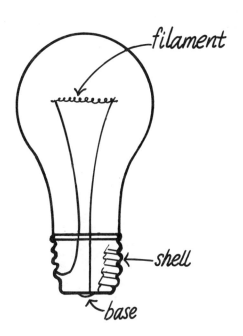

A filament is a long, thin, coiled wire usually made of tungsten, a metal that can be heated to high temperatures. The filament gives off a lot of heat as light is produced. The oxygen has been removed from the bulb because it would cause the heated filament to burn up quickly.

THREE-WAY BULBS

Notice how three-way bulbs are labeled. Do you see a pattern?

30 watts-70 watts-100 watts

50 watts-100 watts-150 watts

100 watts-150 watts-_____watts

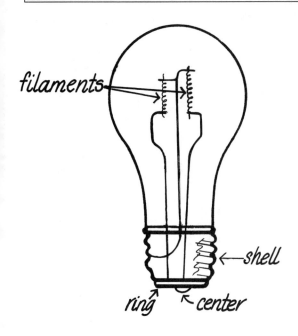

A three-way light bulb has two filaments and three contact points-at the center, ring, and shell. The lower-wattage filament is connected to the ring and shell contacts, the higher-wattage filament to the center and shell contacts. When the switch is turned to the highest wattage position, a current runs through both circuits. The sum of the first two wattages equals the third wattage.

Using three colors, trace the current paths for the lowest wattage, the middle wattage, and the highest wattage. What part of the wire is shared by both circuits?

Glowing Report

5. Make a class bar graph to show total watts for each student's house.

6. Write two true statements from the data on the graph.

_____

_____

_____

_____

Scenarios for lighting design

1. Mr. and Mrs. Alvarez and their three children, Alicia, Gilbert, and Juan, live in a four-bedroom, two-bath home. They cannot afford more than 2,700 watts in lighting. Juan, the youngest, is scared of the dark.

2. Your older sister got married last year. She and her husband are renting a two-bedroom, two-bath condo. They are trying to save money to buy a house. Design lighting with 2,300 watts or less.

3. The Lee family, two parents and four children, share a three-bedroom, two-bath home. You may use up to 3,000 watts for lighting.

4. Four college students are sharing a two-bedroom, two-bath apartment. They spend a lot of time studying and on personal grooming. Their lighting should not exceed 2,400 watts.

5. Your grandparents live in a two-bedroom, one-bath house. They need good lighting to read. They must have lighting with no more than 2,100 watts.

6. Mrs. Morgan repairs and alters clothes in her three-bedroom, two-bath home. She needs good lighting by her sewing machine and ironing board. Her 11-year-old daughter, Allison, lives with her. Design lighting with 2,900 watts or less.

7. A single parent with two teen-agers lives in a two-bedroom, one-bath apartment. How would you light their place so it uses no more than 2,200 watts?

8. Mr. Dawson, an artist who works at home, needs good lighting in his work space. He rents a two-bedroom, 1-bath home. He can have up to 2,100 watts of lighting.

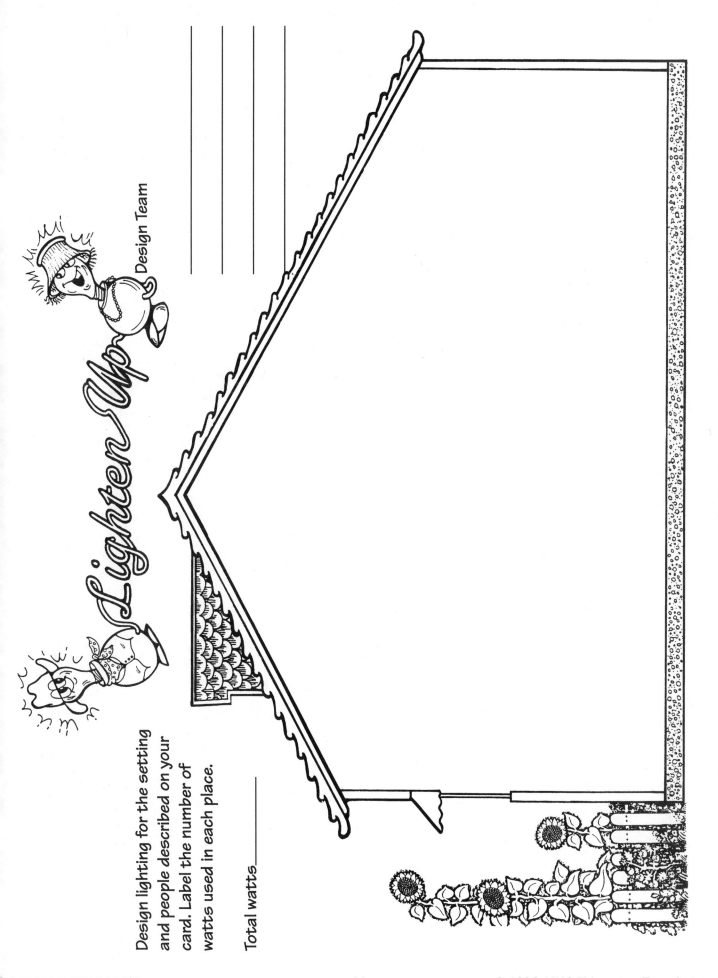

Lighten Up

Design Team _____

Design lighting for the setting and people described on your card. Label the number of watts used in each place.

Total watts _____

Which is Better?

INCANDESCENT BULB

How it works: Current travels through a circuit, heating the filament to a white hot glow. Light is produced by heat.

Energy Use: 5-8% of energy produced is converted into light

FLUORESCENT BULB

How it works: A ballast heats electrodes which release free electrons. Electrons cause vapors to give off ultraviolet rays which make the bulb's phosphorus coating glow.

Energy Use: 13-15% of energy produced is converted into light

Approximate Lumens	Incandescent Bulb		Fluorescent Bulb
900 lumens	60 watts	=	15 watts
1200 lumens	75 watts	=	20 watts
1500 lumens	90 watts	=	26 watts

Cost: _____ per bulb

Average Life: 750-1,000 hours

Cost: _____ per bulb

Average Life: 8,000-10,000 hours

Construct a bar graph showing the costs (bulb and energy) for comparable incandescent and fluorescent bulbs over 9,000 hours of use.

☐ Cost of bulbs

☐ Energy cost

_____
_____
_____
_____
_____
_____
_____
_____

$0 _____

   incandescent   fluorescent

  _____ watts   _____ watts

Kwh = (watts x hours)/1,000
Utility company rate per kwh: _____ ¢

$STATIC$ $MAGIC$

I. Topic Area
Static electricity

II. Introductory Statement
The student will have a variety of experiences with static electricity; discover what happens to materials when he makes and uses static electricity; records his findings on the work sheets.

III. Math Skills
a. Graphs
b. Estimating
c. Measurement

Science Processes
a. Observing and classifying
b. Predicting and hypothesizing
c. Gathering and recording data
d. Interpreting data
e. Applying and generalizing

IV. Materials
Each group will have:
3 shakers of salt
3 shakers of pepper
3 sheets of colored paper
3 pieces of plastic
3 old nylon stockings
3 pieces of cotton material
3 pieces of wool material
3 balloons
Metric ruler
Clay

V. Key Question
Which material carries the strongest electrical charge?

VI. Background
When something has an electric charge, it is said to be electrically charged. Friction can cause some things to become electrically charged. Objects can become electrically charged by adding or taking away electrons. Static electricity is produced by rubbing two different kinds of material together. Static electricity attracts things to it. Static means "not moving." Sometimes static electric charges or static electricity is called friction electricity. You will discover that charges stay on plastic, rubber, and nylon for a long time.

VII. Management
Two days of 45 minute class sessions. Divide the class into small groups of five to six students. On the tables place the materials needed for the two activities. Place the work sheets near each set of materials.

VIII. Procedure
Tell the students that they can have fun with static electricity (pretend that you are a magic person and show how salt and pepper separate). Encourage the student to ask questions about doing the activities. Have the student look at their work sheets. Students will predict what will happen. They need to verify their predictions on the work sheets after doing the activity.

Mix salt and pepper together on a sheet of paper and spread it evenly. Rub a piece of plastic quickly two or three times and hold it above the salt and pepper. Students should observe that the grains of pepper "jump" to the plastic because pepper is lighter in weight than salt. The pepper should stick to the plastic.

After all the students have investigated on their own, discuss the activity and record data on the graph on the work sheet.

1. What happens when you hold the plastic near the salt and pepper?
2. Why do you think the pepper "jumps" to the plastic?

Test your prediction! Mix salt and pepper and spread it evenly on a sheet of paper. Create static electricity by quickly rubbing the material you are testing with your hand. Place metric ruler in clay to stand it up. Hold it 10 cm. above the salt and pepper. If the pepper is attracted to the material, note the height by recording on the graph. If the pepper is not attracted to the material, renew the charge and test at a lower height. Keep testing until you reach the level of attraction.

IX. What the Students Will Do
Included in procedures.

X. Discussion
1. Do you think you can use static electricity to separate pepper from salt?
2. How can you find out?
3. How will you make static electricity?
4. How can you charge a nylon stocking with electricity?
5. How can you charge the wool material with electricity?
6. How can you charge the balloon with electricity?

XI. **Extensions**

Experiment with static electricity. Have children try this experiment on different kinds of days, such as dry, sunny days; rainy days; cool, cloudy days and keep records of their observations. If the classroom is carpeted, or has a rug on the floor, children can do the experiment at school. If not, suggest that they do it at home. Suggest that the children shuffle across the rug on each kind of day and touch a metal object such as a doorknob.

1. What did you feel?
2. What did you hear?
3. If the room is darkened, what do you see?
4. What makes the spark?

XII. **Curriculum Coordinates**

Books

1. Bill, Thelma Harrington, *Thunderstorm*, The Viking Press, Inc., New York, 1960.

Films

1. *Electricity All About Us* (11 minutes, color); Coronet Instructional Films. Demonstrates the nature of static and current electricity.

Filmstrip

1. *Experimenting with Static Electricity* (color); Encyclopedia Britannica. Includes experiments children can do with electrically charged objects.

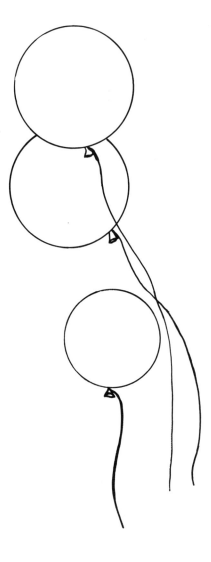

NAME _____

STATIC MAGIC

WHICH MATERIAL CARRIES THE STRONGEST ELECTRICAL CHARGE?

<u>TESTING MATERIALS</u>

1. PLASTIC MATERIAL
2. COTTON MATERIAL
3. WOOL MATERIAL
4. NYLON STOCKING
5. BALLOON

PREDICT BY LISTING THE ITEMS FROM STRONGEST TO WEAKEST ATTRACTION.

PREDICTION

1. _____
2. _____
3. _____
4. _____
5. _____

ACTUAL

1. _____
2. _____
3. _____
4. _____
5. _____

TEST YOUR PREDICTION!

1. MIX SALT AND PEPPER AND SPREAD IT EVENLY ON A SHEET OF PAPER.

2. CREATE STATIC ELECTRICITY BY QUICKLY RUBBING THE MATERIAL YOU ARE TESTING WITH YOUR HAND.

3. PLACE METRIC RULER IN CLAY TO STAND UP. HOLD THE MATERIAL ABOVE THE SALT AND PEPPER STARTING AT 10 CM. IF THE PEPPER IS ATTRACTED TO THE MATERIAL NOTE THE HEIGHT ON THE GRAPH.

4. IF THE PEPPER IS NOT ATTRACTED TO THE TEST MATERIAL, RENEW THE CHARGE AND TEST AT A LOWER HEIGHT. KEEP TESTING UNTIL YOU REACH THE LEVEL OF ATTRACTION.

GRAPH

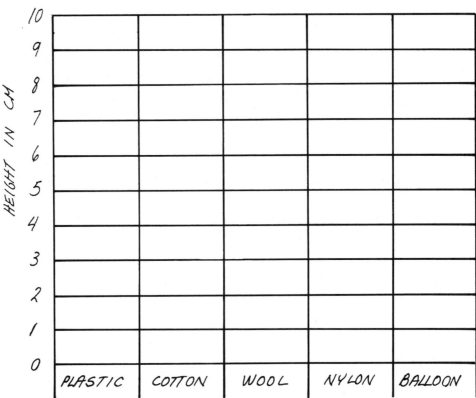

| | PLASTIC | COTTON | WOOL | NYLON | BALLOON |

TYPE OF MATERIAL

(Y-axis label: HEIGHT IN CM, 0–10)

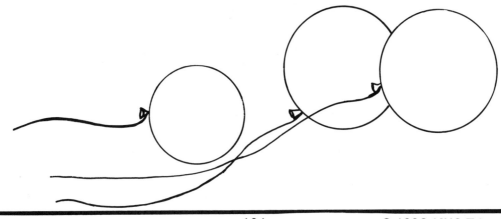

Slip, Sliding Away

Topic
Friction

Key Question
How can we reduce friction?

Focus
Students will slide an object down an inclined plane in order to compare the friction-reducing capabilities of several lubricants.

Guiding Documents
Project 2061 Benchmark
"Students should have lots of experiences to shape their intuition about motion and forces long before encountering laws. Especially helpful are experimentation and discussion of what happens as surfaces become more elastic or more free of friction." (page 88)

NRC Standards
- *Employ simple equipment and tools to gather data and extend the senses.*
- *Communicate investigations and explanations.*
- *The motion of an object can be described by its position, direction of motion, and speed. That motion can be measured and represented on a graph.*

NCTM Standards
- *Make and use measurements in problems and everyday situations*
- *Collect, organize, and describe data*

Math
Measurement
  length
  mass
  time (digital)
Decimals
Average
  median
Graphing
Ordering

Science
Physical science
  force and motion
  friction

Integrated Processes
Observing
Collecting and recording data
Comparing and contrasting
Controlling variables
Interpreting data
Relating

Materials
For the class:
  meter sticks or measuring tapes
  balance and gram masses
  paper/cloth towels, liquid soap, and water for cleanup

For each group:
  desk or table (see *Management 1*)
  books or other props (see *Management 1*)
  film canister filled with sand
  lubricants such as soap, wax, and oil
   (see *Management 2*)
  small piece of sponge (see *Management 3*)
  15 cm by 6 cm piece of corrugated cardboard
  masking tape
  stopwatch
  colored pencils

Background Information
  Friction is a force, a resistance to motion. It is sometimes called a hidden force because it is such a part of our everyday lives that we often overlook it. Whenever two surfaces rub against each other there is sliding friction. The bumps of one surface catch the bumps of the other surface. This is not only true for solid against solid but also for solid against gas such as air. Think of a parachute where air resistance or friction is needed to slow its descent or of an aerodynamically designed car where friction between air and car has been reduced, increasing fuel efficiency.
  The amount of friction varies with the kind of surface. Smooth surfaces tend to create less friction than rough surfaces. But even the smoothest surfaces have slight imperfections which cause friction; no surface is completely free of friction. Consider a playground slide. If the surface of the slide and the clothes of the person coming in contact with the slide are relatively smooth, the ride will be fast.
  The amount of friction also varies with the force (weight) pressing the surfaces together. The greater the mass, the amount of material of an object, the greater the weight and the greater the friction. In this activity, we chose to measure mass because balances are more commonly available than the spring scales used to measure weight.
  Friction produces heat and causes wear and tear on surfaces. Reducing friction can increase the life of machinery. Sometimes friction is reduced to increase speed.
  Friction can be reduced through the use of lubricants. Lubricants smooth out a surface, filling in the nooks and crannies. A dull safety pin coated with soap is easier to

use. Oil and grease smooth the surfaces of bearings and gears in machines. Butter may help remove a ring from a swollen finger or a hand stuck in a jar. Skis are waxed to make them faster on the snow. In the sport of curling, a stone slides across a sheet of ice toward a target as team members sweep the ice with brooms to reduce friction. Desk drawers which stick can be waxed. We spray silicone, a friction-reducer, to cure squeaky door hinges. Water, talc, and graphite (from a soft pencil) can also be used as lubricants.

Sometimes friction is desirable. We want bicycle and car tires to grip the road. Although rubber is a high-friction material, the rough surface created by the tire tread contributes to friction. Friction is particularly needed under rainy conditions where a layer of water may sit on top of a road naturally lubricated by oils from daily traffic. Car brakes utilize friction in order to stop the car. We rely upon the friction created by the soles of our shoes so we don't slide across the basketball court or slip on the sidewalk. Friction caused when striking a match produces a flame. The friction produced from rubbing a bow against a violin string creates music.

Management

1. A flat desk or table at least 50 cm in length is preferred. To transform it into an inclined plane, use books to prop two legs about 10 cm off the floor. Use the sand-filled film canister to do a test slide or two, adjusting the height if the canister slides too quickly or doesn't move at all. The height needed will vary with the length and surface material of the inclined plane and the sliding object used. A safety caution: the desk or table should not be propped so high that it is in danger of falling over.

   Tape the piece of cardboard to the desk as shown. The cardboard keeps the film canister from dropping to the floor and defines when the slide stops.

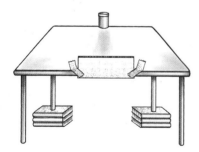

2. After preliminary discussion, collect the lubricants the students want to try. These might include a bar of soap, dishwashing soap mixed with water, wax, vegetable oil, heavy car oil (30- or 40-weight), etc.

3. Whatever lubricant is applied to the desk should also be applied to the bottom of the film canister. A small piece of sponge, about 1-inch by 2-inches, is useful for applying oils. If more than one oil is tested, a different sponge should be used for each. Remove excess oil with a paper towel, leaving a lightly-coated surface. The same should be done with soap, wax, etc. The thickness of the coating is a variable; coats which are too thick can actually increase friction.

4. Times will be very fast, sometimes less than a second. Close observation and good reflexes are important. To control variables, the start and stop of the slide should be defined. The stopwatch should be started when the film canister is released, its back side lined up with the edge of the desk, and stopped when it hits the cardboard barrier.

5. Each group can test the various lubricants on their assigned desk, *thoroughly cleaning* its surface and the film canister before applying the next one. Or set up stations, with one desk and film canister designated for bar soap, another for vegetable oil, etc. and have groups rotate from desk to desk to gather data.

6. Alternate materials: If suitable desks or film canisters are in short supply, large cookie sheets can be used as inclined planes and objects with a mass between 50 and 80 grams, such as an olive can, may be used for sliding.

7. Increments used for the graph will depend on the range of data. If times are all four seconds or less, increments of .20 (two-tenths of a second) can be used.

8. While reading a digital stopwatch is not difficult, interpreting the decimals when constructing a graph can be. Students need to understand where, between 1.80 and 1.90, 1.84 lies along a number line so they will know where to mark the bar on the graph.

The following is offered for those students ready for more independent work.

> Present this *Open-ended Challenge:* You work in a factory. Your product slides down a ramp to get to the next part of the assembly line. Without moving the ramp, how could you increase the sliding speed?
>
> Students should describe the sliding surface and its length, the height of the ramp, and the product and its mass. After data is gathered, the results should be illustrated in some way.

Procedure

Day One: Setting the Stage

1. Ask students to tell about kinds of slides they have used. [playground slides, water park slides, etc.] What makes slides fun? [going fast] What about the slides make them fast? [the smooth metal of a playground slide, the running water going down the water park slide, sitting on a mat to go down a huge plastic slide]

2. Tell students they are going to investigate how to make a slide faster. The more slippery it is, the less friction there is. Explain the "slide" setup that will be used.

3. Have students brainstorm possible lubricants that might reduce friction. From this list, come to a consensus on four or five which are obtainable and would be interesting to test.

4. Encourage students to help bring in these items the next day or gather them yourself.

Day Two: Investigating

1. Instruct students on how to set up the slide and have them complete this task.

2. Distribute the first activity page and have students complete the measurements of the desk and film canister.
3. Tell students they will be recording three times for each lubricant, starting with "none," an untreated surface. Review how timing is to be done (see *Management 4*).
4. Remind students to rub away excess lubricant, leaving only a light coating, and to thoroughly clean the desk and film canister before applying the next lubricant. Have them continue until all the lubricants have been tested.
5. Instruct students to look at their three slide times for "none." Have them use a colored pencil to color the highest time and lowest time. The remaining or middle number is the median average and should be written in the third column. Repeat for each of the other lubricants.

Surface Treatment	Slide Times	Average Time*
none	3.29	
	4.66	3.53
	3.53	
bar soap	2.18	
	1.63	1.84
	1.84	

6. Give students the graph page and discuss together how to number the increments. Once the graph is labeled, have students color the bars according to their data.
7. Direct students to order the lubricants from best (fastest) to worst (slowest).
8. Have students communicate their results and raise new questions they may want to explore.

Discussion
1. How do the lubricants compare at reducing friction? [Example: on a melamine surface, the oils seemed to reduce friction more than soaps] Were there any that increased friction? If so, why do you think so?
2. How do your results compare with those of other groups? What might explain differences? [how much of the lubricant was applied to the surface, accuracy of timing, etc.]
3. What other variables would affect the results? [the kind of surface used (wood, plastic, metal, etc.), the length of the ramp, the height of the ramp, the mass and kind of surface of the sliding object]
4. Do you think different lubricants would perform better on different surfaces? In other words, is what best reduces friction between metals different than what best reduces friction between plastics or wood?
5. Give examples where you would want to reduce friction in the real world. [all kinds of machinery, slides, skis and sleds on snow, skates on ice, etc.]

6. When would you want to have some friction? [shoes, bicycle or car tires, conveyor belts, to stop roller blades, etc.]

Extensions
1. Run tests comparing different weights of car oils, 10-weight to 40-weight. How is the recommended oil weight related to temperature? [Low weights are thinner and used in cold weather while high weights are thicker and used in warm weather. Oils with a variable range of weight, such as 10-30, can also be purchased. The consistency of the oil changes with the temperature.]
2. Compare the performance of lubricants on different surfaces such as metal, wood, plastic, or some other material. Use a metal sliding object with a metal inclined plane, a wooden sliding object with a wooden inclined plane, etc.

107

Slip, Sliding Away

How can we reduce friction?

Lubricant	Slide Times		Average Times*

*Color the high and low times for each lubricant. Write the remaining number, the median average, in the last column.

Inclined Plane

Surface:

Length:

Height:

Sliding Object

Kind:

Mass:

Slip, Sliding Away

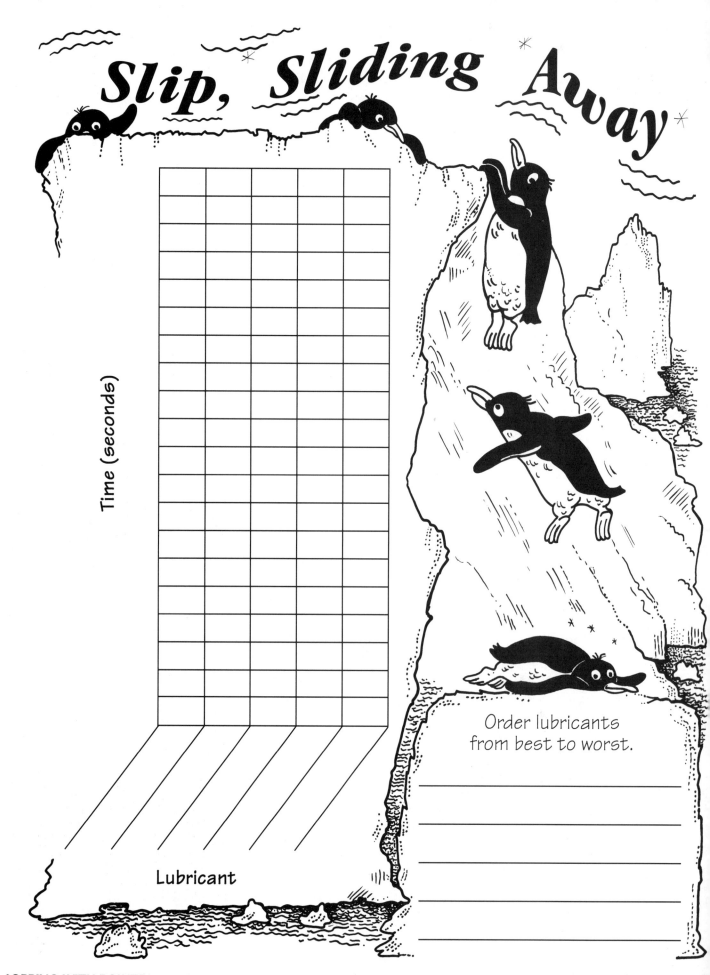

Time (seconds)

Lubricant

Order lubricants from best to worst.

_____

_____

_____

_____

_____

Catapults

Topic
Testing variables via catapults

Key Question
What can you discover about catapults?

Focus
Students will construct a catapult, experiment with accuracy and distance challenges, and run controlled tests of variables which influence how far or high a small sponge can be flung.

Guiding Documents
Project 2061 Benchmarks
- *Measuring instruments can be used to gather accurate information for making scientific comparisons of objects and events and for designing and constructing things that will work properly.*
- *Something that is moving may move steadily or change its direction. The greater the force is, the greater the change in motion will be. The more massive an object is, the less effect a given force will have.*
- *Recognize when comparisons might not be fair because some conditions are not kept the same.*

NRC Standards
- *Scientists use different kinds of investigations depending on the questions they are trying to answer. Types of investigations include describing objects, events, and organisms; classifying them; and doing a fair test (experimenting).*
- *The position and motion of objects can be changed by pushing or pulling. The size of the change is related to the strength of the push or pull.*

NCTM Standards
- *Make and use measurements in problems and everyday situations*
- *Collect, organize, and describe data*

Math
Estimation
Measurement
   length
Statistics
   median average
Graphing

Science
Physical science
   force and motion

Integrated Processes
Observing
Estimating
Collecting and recording data
Comparing and contrasting
Identifying and controlling variables
Interpreting data
Relating

Materials
For each group:
   2" x 4" piece of wood about 8" long
     (or minimum 1" x 3 1/2" x 7")
   two 8-penny nails (2 1/2" long)
   scissors
   1 tongue depressor
   5 toothpicks
   glue
   1 rubber band (1/8" x 2")
   2-cm square piece of sponge
   meter stick or long meter tape

For the class:
   several hammers
   cloth duct tape
   newspaper
   masking tape
   4 bright colors of construction paper
   chart or butcher paper
   small paper clips
   string

Background Information
Catapults provide a context for exploring variables, the focus of this activity. As students modify and tinker with the catapults during the challenges, they will discover what kinds of changes or variables affect the flight of the sponge. This playful experience then leads to controlled tests. A controlled or fair test isolates one variable only; all the other variables are kept the same. Whatever changes then occur in the flight of the sponge can be attributed to how this one variable was changed. If two variables were uncontrolled, it would be difficult to determine which caused differences in the results.

Catapults also illustrate potential and kinetic energy. When triggered, the potential energy stored in the stretched rubber band is transferred by the arm of the catapult to the sponge. This released potential energy is quickly converted to the kinetic energy of the moving sponge. As students move the rubber band higher and bend the arm of the catapult against it, they will be able to feel the increased tension, the increased potential energy.

Management

1. Groups of three or four should construct and test the catapults.
2. The activity will take several days. One plan might proceed as follows:
   - First day – construct the catapults
   - Second day – explore one or more challenges
   - Third day – test the rubber band height variable
   - Fourth day – test a variable of their choice
3. A large area, preferably indoors, will be needed to accomodate all the groups. To test the catapults, each group will need a minimal length of 6 meters. The activity can be done outdoors on a calm day.
4. Safety should be strongly emphasized, both during hammering and when using the catapults. Consider using safety goggles to protect eyes. If you prefer, you can hammer the nails into each piece of wood beforehand.
5. Prepare several compass drawing tools by tying a small paper clip to each end of a piece of string so that the total length from the tip of one paper clip to the tip of the other is 30 cm. Each group can have their own or take turns using them as they become available.

6. Cut two-centimeter squares of construction paper in four bright colors. Each group will need ten squares of each color.

Procedure

Constructing the catapults

1. Give students the construction page and the necessary materials. Have them follow the directions on the page. Do not use the catapults until the glue is completely dry, a half day or more.

Accuracy challenge

2. Distribute the challenge page. To prepare for the accuracy challenge, have each group use the string compass to draw a 60-cm circle on chart paper.
3. Instruct students to cut out the circle, mark their launch line with tape or by other means, and place the center of the circle three meters from the launch line.

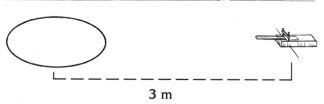

3 m

4. Direct students to experiment with their catapults until they have found a consistent way to land the sponge in the circle. Remind students that, to safely

conduct a launch, group members should be clear of the launch area and be aware of when a launch is to be made.
5. Have students conduct ten launches and record the results as well as explain the setup standards (how they controlled variables) they used to achieve their results.
6. Compare and contrast as a class.

Distance challenge

7. Have each group experiment with their catapult to obtain the longest distance possible, recording their setup standards as well as the results.
8. Compare and contrast as a class.
9. "When you were doing the challenges, you tried different methods for launching. What were some of the things you changed? [how far back the arm of the catapult was pulled, the position of the sponge along the arm, the height of the rubber band, the angle of the wood base] These are called variables, something you can change that affects the results."

Testing variables

10. Inform students that today they will be doing a controlled experiment to test one variable, the height of the rubber band.
11. Distribute the third activity page. Have students write their predictions. Tell them that the other variables will be controlled as follows: the arm of the catapult will be pulled all the way back to the board, the board will lie flat on the ground, and the sponge will be placed right behind the toothpick on top of the arm.
12. Have students make a newspaper landing path about six meters long by taping newspapers together and marking a starting line. Give each group ten squares in each of four colors of construction paper and some glue. These materials plus the catapult, the sponge, and a meter stick will be taken to the launch site.
13. Take the class to the launch site and assign an area to each group. Tell each group to lay out their landing path and position the rubber band on the nails so that it rests just above the 1-cm toothpick mark on the launcher.

14. Inform the class that they will glue the _____ (name color) squares to mark the landing place for the ten launches at the 1-cm mark. Mark the landing with an x first, then glue a colored square in place.
15. Ask each group to assign jobs to its members: launcher, retriever, marker/measurer, and square-gluer. They will rotate jobs each time the rubber band position is changed.

16. Have the groups perform the ten launches.
17. Direct the groups to move the rubber band to the next height, use a different-color square to mark landings, and make ten launches.
18. Repeat the procedure for the other heights.
19. Instruct the marker in each group to draw a line between the distances of the 5th and 6th squares from the launch line in each set of colors. This median distance should be measured, recorded, and graphed on the activity sheet.

median distance

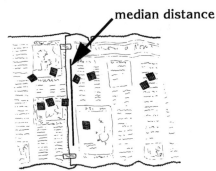

20. Discuss the results.
21. Have students set up and perform a test of another variable.

Discussion
1. What things did you change, or experiment with, as you worked on the challenge?
2. To make a fair test, we need to control all but one variable. How did you make a fair test?
3. Describe how the sponge traveled through the air at lower rubber band heights compared to higher rubber band heights. [the arc gets higher as the rubber band height is increased]
4. What other things did you discover about the catapults?

Extension
Explore other variables such as the mass of the object being launched (various sizes of sponge or other materials) and different sizes of rubber bands.

Curriculum Correlation
Social Science
Have students research the construction and historic use of catapults. A *Project 2061 Benchmark* states: *Throughout all of history, people everywhere have invented and used tools. Most tools of today are different from those of the past but many are modifications of very ancient tools.*

Catapults

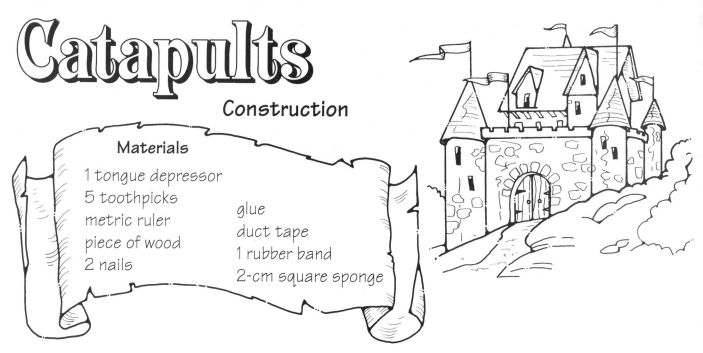

Construction

Materials

1 tongue depressor
5 toothpicks
metric ruler
piece of wood
2 nails

glue
duct tape
1 rubber band
2-cm square sponge

1. For the arm of the catapult, cut one tip of the tongue depressor to make a straight edge. Mark 1, 2, 3, and 4 cm lines from this edge. Glue pieces of toothpick along each line. On the other side, glue a piece of toothpick 3 cm from the rounded tip. Let dry.

2. Draw a line across the width of the board about 6 cm from the end. Hammer two nails along this line, 6.5 cm apart.

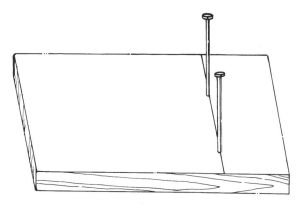

3. Assemble the catapult with duct tape. Stretch the rubber band around the nails.

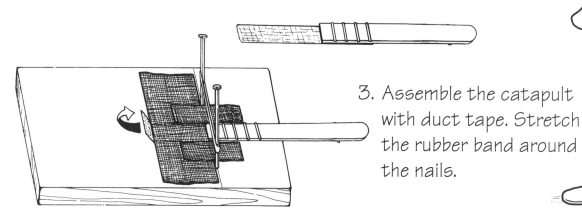

Catapults

Accuracy Challenge

Consistently land the sponge in a 60-cm circle whose center is 3 meters from the catapult.

Explain your launch setup.

Of ten launches, how many landed in the circle?

Distance Challenge

Launch the sponge as far as possible.

Explain your launch setup.

Farthest distance:

What changes did you make in trying to meet the challenges?

Catapults

How does the height of the rubber band affect the distance the sponge travels?

Prediction:

How will other variables be controlled?

Make ten launches at each height.

Height of Rubber Band	Average Distance (median)*
1 cm	
2 cm	
3 cm	
4 cm	

* Measure a mark made between the 5th and 6th landings. This is the median, or middle, distance.

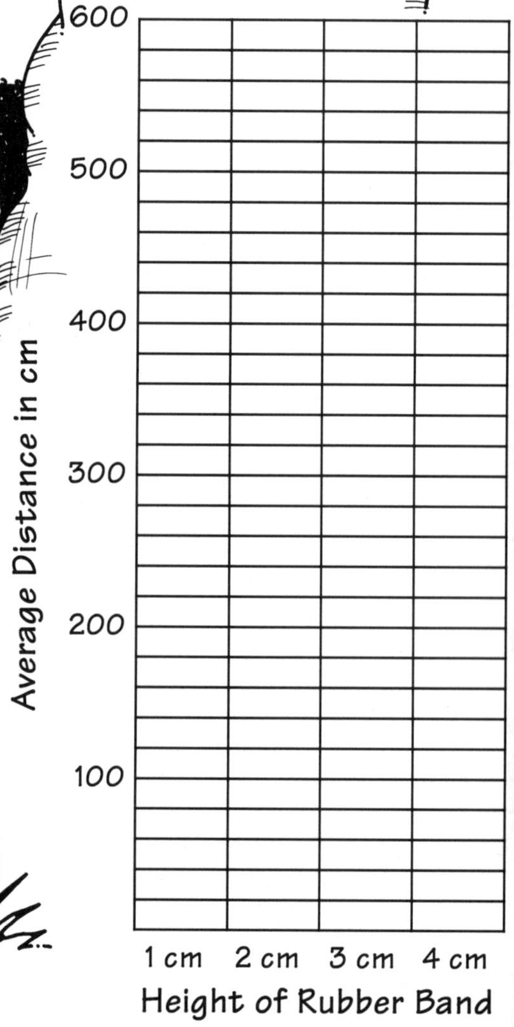

What did you observe?

How did your prediction compare with the results?

Average Distance in cm

600
500
400
300
200
100

1 cm   2 cm   3 cm   4 cm
Height of Rubber Band

© 1996 AIMS Education Foundatio

Catapults

Variable to be tested:

How will other variables be controlled?

Record and graph your results.

Conclusions:

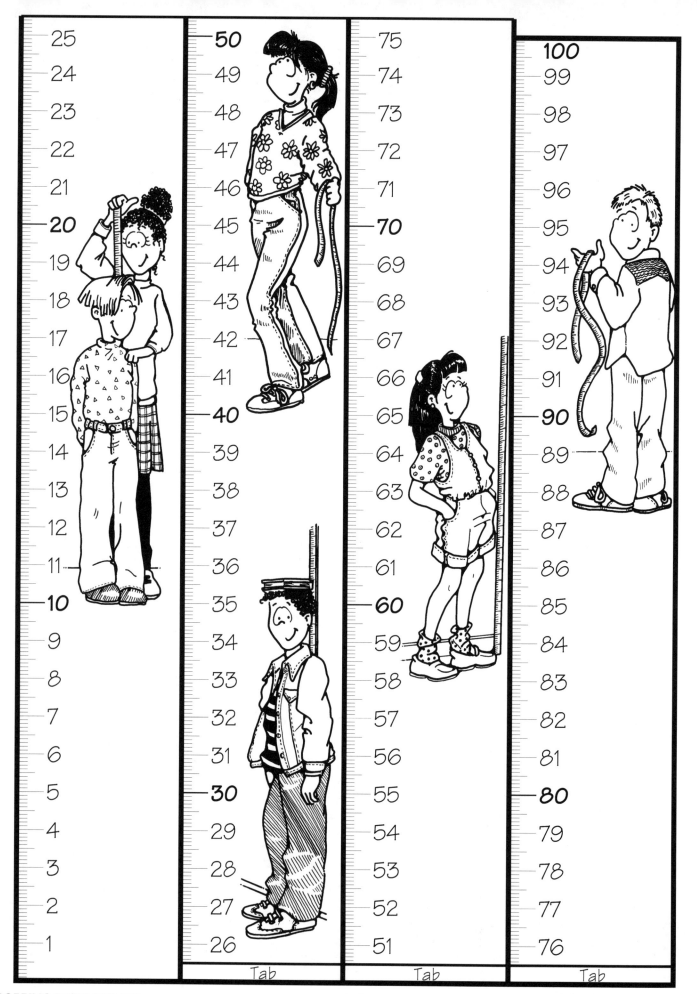

25	50	75	100
24	49	74	99
23	48	73	98
22	47	72	97
21	46	71	96
20	45	70	95
19	44	69	94
18	43	68	93
17	42	67	92
16	41	66	91
15	40	65	90
14	39	64	89
13	38	63	88
12	37	62	87
11	36	61	86
10	35	60	85
9	34	59	84
8	33	58	83
7	32	57	82
6	31	56	81
5	30	55	80
4	29	54	79
3	28	53	78
2	27	52	77
1	26	51	76
	Tab	Tab	Tab

The AIMS Program

AIMS is the acronym for "**A**ctivities **I**ntegrating **M**athematics and **S**cience." Such integration enriches learning and makes it meaningful and holistic. AIMS began as a project of Fresno Pacific University to integrate the study of mathematics and science in grades K-9, but has since expanded to include language arts, social studies, and other disciplines.

AIMS is a continuing program of the non-profit AIMS Education Foundation. It had its inception in a National Science Foundation funded program whose purpose was to explore the effectiveness of integrating mathematics and science. The project directors in cooperation with 80 elementary classroom teachers devoted two years to a thorough field-testing of the results and implications of integration.

The approach met with such positive results that the decision was made to launch a program to create instructional materials incorporating this concept. Despite the fact that thoughtful educators have long recommended an integrative approach, very little appropriate material was available in 1981 when the project began. A series of writing projects have ensued and today the AIMS Education Foundation is committed to continue the creation of new integrated activities on a permanent basis.

The AIMS program is funded through the sale of this developing series of books and proceeds from the Foundation's endowment. All net income from program and products flows into a trust fund administered by the AIMS Education Foundation. Use of these funds is restricted to support of research, development, and publication of new materials. Writers donate all their rights to the Foundation to support its on-going program. No royalties are paid to the writers.

The rationale for integration lies in the fact that science, mathematics, language arts, social studies, etc., are integrally interwoven in the real world from which it follows that they should be similarly treated in the classroom where we are preparing students to live in that world. Teachers who use the AIMS program give enthusiastic endorsement to the effectiveness of this approach.

Science encompasses the art of questioning, investigating, hypothesizing, discovering, and communicating. Mathematics is the language that provides clarity, objectivity, and understanding. The language arts provide us powerful tools of communication. Many of the major contemporary societal issues stem from advancements in science and must be studied in the context of the social sciences. Therefore, it is timely that all of us take seriously a more holistic mode of educating our students. This goal motivates all who are associated with the AIMS Program. We invite you to join us in this effort.

Meaningful integration of knowledge is a major recommendation coming from the nation's professional science and mathematics associations. The American Association for the Advancement of Science in *Science for All Americans* strongly recommends the integration of mathematics, science, and technology. The National Council of Teachers of Mathematics places strong emphasis on applications of mathematics such as are found in science investigations. AIMS is fully aligned with these recommendations.

Extensive field testing of AIMS investigations confirms these beneficial results.

1. Mathematics becomes more meaningful, hence more useful, when it is applied to situations that interest students.
2. The extent to which science is studied and understood is increased, with a significant economy of time, when mathematics and science are integrated.
3. There is improved quality of learning and retention, supporting the thesis that learning which is meaningful and relevant is more effective.
4. Motivation and involvement are increased dramatically as students investigate real-world situations and participate actively in the process.

We invite you to become part of this classroom teacher movement by using an integrated approach to learning and sharing any suggestions you may have. The AIMS Program welcomes you!

AIMS Education Foundation Programs

A Day with AIMS

Intensive one-day workshops are offered to introduce educators to the philosophy and rationale of AIMS. Participants will discuss the methodology of AIMS and the strategies by which AIMS principles may be incorporated into curriculum. Each participant will take part in a variety of hands-on AIMS investigations to gain an understanding of such aspects as the scientific/mathematical content, classroom management, and connections with other curricular areas. *A Day with AIMS* workshops may be offered anywhere in the United States. Necessary supplies and take-home materials are usually included in the enrollment fee.

A Week with AIMS

Throughout the nation, AIMS offers many one-week workshops each year, usually in the summer. Each workshop lasts five days and includes at least 30 hours of AIMS hands-on instruction. Participants are grouped according to the grade level(s) in which they are interested. Instructors are members of the AIMS Instructional Leadership Network. Supplies for the activities and a generous supply of take-home materials are included in the enrollment fee. Sites are selected on the basis of applications submitted by educational organizations. If chosen to host a workshop, the host agency agrees to provide specified facilities and cooperate in the promotion of the workshop. The AIMS Education Foundation supplies workshop materials as well as the travel, housing, and meals for instructors.

AIMS One-Week Perspectives Workshops

Each summer, Fresno Pacific University offers AIMS one-week workshops on its campus in Fresno, California. AIMS Program Directors and highly qualified members of the AIMS National Leadership Network serve as instructors.

The Science Festival and the Festival of Mathematics

Each summer, Fresno Pacific University offers a Science Festival and a Festival of Mathematics. These festivals have gained national recognition as inspiring and challenging experiences, giving unique opportunities to experience hands-on mathematics and science in topical and grade-level groups. Guest faculty includes some of the nation's most highly regarded mathematics and science educators. Supplies and take-home materials are included in the enrollment fee.

The AIMS Instructional Leadership Program

This is an AIMS staff-development program seeking to prepare facilitators for leadership roles in science/math education in their home districts or regions. Upon successful completion of the program, trained facilitators may become members of the AIMS Instructional Leadership Network, qualified to conduct AIMS workshops, teach AIMS in-service courses for college credit, and serve as AIMS consultants. Intensive training is provided in mathematics, science, process and thinking skills, workshop management, and other relevant topics.

College Credit and Grants

Those who participate in workshops may often qualify for college credit. If the workshop takes place on the campus of Fresno Pacific University, that institution may grant appropriate credit. If the workshop takes place off-campus, arrangements can sometimes be made for credit to be granted by another college or university. In addition, the applicant's home school district is often willing to grant in-service or professional development credit. Many educators who participate in AIMS workshops are recipients of various types of educational grants, either local or national. Nationally known foundations and funding agencies have long recognized the value of AIMS mathematics and science workshops to educators. The AIMS Education Foundation encourages educators interested in attending or hosting workshops to explore the possibilities suggested above. Although the Foundation strongly supports such interest, it reminds applicants that they have the primary responsibility for fulfilling *current* requirements.

For current information regarding the programs described above, please complete the following:

Information Request

Please send current information on the items checked:

___ *Basic Information Packet* on AIMS materials
___ *Festival of Mathematics*
___ *Science Festival*
___ *AIMS Instructional Leadership Program*

___ *AIMS One-Week Perspectives* workshops
___ *A Week with AIMS* workshops
___ Hosting information for *A Day with AIMS* workshops
___ Hosting information for *A Week with AIMS* workshops

Name _____ Phone _____

Address _____

Street   City   State   Zip

AIMS Program Publications

GRADES K-4 SERIES

Bats Incredible
Brinca de Alegria Hacia la Primavera con las Matemáticas y Ciencias
Cáete de Gusto Hacia el Otoño con la Matemáticas y Ciencias
Cycles of Knowing and Growing
Fall Into Math and Science
Field Detectives
Glide Into Winter With Math and Science
Hardhatting in a Geo-World (Revised Edition, 1996)
Jaw Breakers and Heart Thumpers (Revised Edition, 1995)
Los Cincos Sentidos
Overhead and Underfoot (Revised Edition, 1994)
Patine al Invierno con Matemáticas y Ciencias
Popping With Power (Revised Edition, 1996)
Primariamente Física (Revised Edition, 1994)
Primarily Earth
Primariamente Plantas
Primarily Physics (Revised Edition, 1994)
Primarily Plants
Sense-able Science
Spring Into Math and Science
Under Construction

GRADES K-6 SERIES

Budding Botanist
Critters
El Botanista Principiante
Exploring Environments
Mostly Magnets
Ositos Nada Más
Primarily Bears
Principalmente Imanes
Water Precious Water

GRADES 5-9 SERIES

Actions with Fractions
Brick Layers
Conexiones Eléctricas
Down to Earth
Electrical Connections
Finding Your Bearings (Revised Edition, 1996)
Floaters and Sinkers (Revised Edition, 1995)
From Head to Toe
Fun With Foods
Gravity Rules!
Historical Connections in Mathematics, Volume I
Historical Connections in Mathematics, Volume II
Historical Connections in Mathematics, Volume III
Machine Shop
Magnificent Microworld Adventures
Math + Science, A Solution
Off the Wall Science: A Poster Series Revisited
Our Wonderful World
Out of This World (Revised Edition, 1994)
Pieces and Patterns, A Patchwork in Math and Science
Piezas y Diseños, un Mosaic de Matemáticas y Ciencias
Soap Films and Bubbles
Spatial Visualization
The Sky's the Limit (Revised Edition, 1994)
The Amazing Circle, Volume 1
Through the Eyes of the Explorers:
    Minds-on Math & Mapping
What's Next, Volume 1
What's Next, Volume 2
What's Next, Volume 3

For further information write to:
AIMS Education Foundation • P.O. Box 8120 • Fresno, California 93747-8120

We invite you to subscribe to *AIMS*!

Each issue of *AIMS* contains a variety of material useful to educators at all grade levels. Feature articles of lasting value deal with topics such as mathematical or science concepts, curriculum, assessment, the teaching of process skills, and historical background. Several of the latest AIMS math/ science investigations are always included, along with their reproducible activity sheets. As needs direct and space allows, various issues contain news of current developments, such as workshop schedules, activities of the AIMS Instructional Leadership Network, and announcements of upcoming publications.

AIMS is published monthly, August through May. Subscriptions are on an annual basis only. A subscription entered at any time will begin with the next issue, but will also include the previous issues of that volume. Readers have preferred this arrangement because articles and activities within an annual volume are often interrelated.

Please note that an *AIMS* subscription automatically includes duplication rights for one school site for all issues included in the subscription. Many schools build cost-effective library resources with their subscriptions.

YES! I am interested in subscribing to *AIMS*.

Name _____   Home Phone _____

Address _____   City, State, Zip _____

Please send the following volumes (subject to availability):

_____	Volume V	(1990-91)	$30.00	_____ Volume X	(1995-96)	$30.00
_____	Volume VI	(1991-92)	$30.00	_____ Volume XI	(1996-97)	$30.00
_____	Volume VII	(1992-93)	$30.00	_____ Volume XII	(1997-98)	$30.00
_____	Volume IX	(1994-95)	$30.00	_____ Volume XIII	(1998-99)	$30.00

_____ **Limited offer: Volumes XIII & XIV (1998-2000) $55.00**
(Note: Prices may change without notice)

Check your method of payment:

❏ Check enclosed in the amount of $ _____

❏ Purchase order attached (Please include the P.O.#, the authorizing signature, and position of the authorizing person.)

❏ Credit Card   ❏ Visa   ❏ MasterCard   Amount $ _____

Card # _____   Expiration Date _____

Signature _____   Today's Date _____

Make checks payable to **AIMS Education Foundation.**
Mail to *AIMS* Magazine, P.O. Box 8120, Fresno, CA 93747-8120.
Phone (559) 255-4094 or (888) 733-2467 FAX (559) 255-6396
AIMS Homepage: http://www.AIMSedu.org/

AIMS Duplication Rights Program

AIMS has received many requests from school districts for the purchase of unlimited duplication rights to AIMS materials. In response, the AIMS Education Foundation has formulated the program outlined below. There is a built-in flexibility which, we trust, will provide for those who use AIMS materials extensively to purchase such rights for either individual activities or entire books.

It is the goal of the AIMS Education Foundation to make its materials and programs available at reasonable cost. All income from the sale of publications and duplication rights is used to support AIMS programs; hence, strict adherence to regulations governing duplication is essential. Duplication of AIMS materials beyond limits set by copyright laws and those specified below is strictly forbidden.

Limited Duplication Rights

Any purchaser of an AIMS book may make up to *200 copies* of any activity in that book for use at *one school site.* Beyond that, rights must be purchased according to the appropriate category.

Unlimited Duplication Rights for Single Activities

An individual or school may purchase the right to make an unlimited number of copies of a single activity. The royalty is $5.00 per activity per school site.

Examples: 3 activities x 1 site x $5.00 = $15.00
                     9 activities x 3 sites x $5.00 = $135.00

Unlimited Duplication Rights for Entire Books

A school or district may purchase the right to make an unlimited number of copies of a single, *specified* book. The royalty is $20.00 per book per school site. This is in addition to the cost of the book.

Examples: 5 books x 1 site   x $20.00 = $100.00
                    12 books x 10 sites x $20.00 = $2400.00

Magazine/Newsletter Duplication Rights

Those who purchase *AIMS* (magazine)/*Newsletter* are hereby granted permission to make up to 200 copies of any portion of it, provided these copies will be used for educational purposes.

Workshop Instructors' Duplication Rights

Workshop instructors may distribute to registered workshop participants a maximum of 100 copies of any article and/or 100 copies of no more than eight activities, provided these six conditions are met:

1. Since all AIMS activities are based upon the *AIMS Model of Mathematics* and the *AIMS Model of Learning*, leaders must include in their presentations an explanation of these two models.
2. Workshop instructors must relate the AIMS activities presented to these basic explanations of the AIMS philosophy of education.
3. The copyright notice must appear on all materials distributed.
4. Instructors must provide information enabling participants to order books and magazines from the Foundation.
5. Instructors must inform participants of their limited duplication rights as outlined below.
6. Only student pages may be duplicated.

Written permission must be obtained for duplication beyond the limits listed above. Additional royalty payments may be required.

Workshop Participants' Rights

Those enrolled in workshops in which AIMS student activity sheets are distributed may duplicate a maximum of 35 copies or enough to use the lessons one time with one class, whichever is less. Beyond that, rights must be purchased according to the appropriate category.

Application for Duplication Rights

The purchasing agency or individual must clearly specify the following:
1. Name, address, and telephone number
2. Titles of the books for Unlimited Duplication Rights contracts
3. Titles of activities for Unlimited Duplication Rights contracts
4. Names and addresses of school sites for which duplication rights are being purchased.

NOTE: Books to be duplicated must be purchased separately and are not included in the contract for Unlimited Duplication Rights.

The requested duplication rights are automatically authorized when proper payment is received, although a *Certificate of Duplication Rights* will be issued when the application is processed.

Address all correspondence to: **Contract Division**
AIMS Education Foundation
P.O. Box 8120
Fresno, CA 93747-8120